TRANSFIGURATION

Notes Toward a Radical Catholic Reimagination of Everything

Michael Martin

TRANSFIGURATION

Notes Toward a Radical Catholic Reimagination of Everything

Kevin Dean,
Quarr Abbey I. of W. Nov 2019,

Angelico Press

First published in English by Angelico Press, 2018

For information, address:
Angelico Press
169 Monitor St.
Brooklyn, NY 11222
www.angelicopress.com

978-1-62138-423-6 (pbk)
978-1-62138-424-3 (cloth)
978-1-62138-425-0 (ebook)

Cover image: untitled, by Emi Shigeno,
acrylic on canvas
Cover design: Michael Schrauzer

For Brendan, Dylan, Thomas, Mae,
Aidan, Zelie, Isabel, Gabriel, and Daniel

CONTENTS

ACKNOWLEDGMENTS

MY deepest appreciation goes to Therese Schroeder-Sheker and James Wetmore, two dear friends who encouraged me in this work and read the manuscript in its infancy. Onward! My thanks also to John Riess, who told me I was not crazy. And all of my love to Bonnie.

INTRODUCTION

THIS book arose out of a sense of vocation. It is not a book I wanted to write, but one I felt a responsibility to write. Furthermore, I do not expect that the ideas in this book will be taken up by my contemporaries. That may seem a maudlin confession, but the sense of vocation out of which this book was written did not come with a promise of reward. My only hope is to leave an artifact, a broken shard of pottery from a vessel that once held wine. It no longer belongs to me.

Stella Matutina Farm
The Feast of St. Elijah, 2018

1

A DELICATE EMPIRICISM

GOETHE, SOPHIOLOGY, AND THE POSSIBILITIES OF A CATHOLIC SCIENCE

And science itself, our science—indeed, what is the significance of all science, viewed as a symptom of life? For what—worse yet, *whence*—all science? How now? Is the resolve to be so scientific about everything perhaps a kind of fear of, an escape from, pessimism? A subtle last resort against—*truth*? And, morally speaking, a sort of cowardice and falseness? Amorally speaking, a ruse?

FRIEDRICH NIETZSCHE, "Attempt at a Self-Criticism" (1886)[1]

If one considers this development of modern scientific research, must not one then say to oneself that scientific research is entangling itself in a kind of web, and only outside of this web can one find the world?

RUDOLF STEINER, *The Boundaries of Natural Science* (1920)[2]

1. Friedrich Nietzsche, "Attempt at a Self-Criticism," in *Basic Writings of Nietzsche*, trans. and ed. Walter Kaufmann (New York: The Modern Library, 1992), 18.
2. Rudolf Steiner, *The Boundaries of Natural Science: Eight Lectures* (Spring Valley, NY: Anthroposophic Press, 1983), 8.

WE live in times charged by the political. This is true in the arts and sciences no less than economics and the sphere of rights. It is also a quality that permeates religion. This may be due to what seems to be a deeply held human need to form allegiances, to belong, to find protection from a hostile environment. As Wilhelm Reich observed, the inhabitants of that most hostile of environments, the desert, "armor" themselves (or are armored by evolutionary processes), and their forms—as in lizards and cacti, for example—are characterized by sharp spines, hardened skin surfaces and cortices, and a general inhospitality to touch. Human beings raised in hostile psychological environments, thought Reich, likewise armor themselves.[3] The propensity to turn to the political, then, may indeed be a psychological reflex; that is, a pathology. The political (or, more properly, ideology), becomes in this way the persona, the mask through which one faces the world and behind which one hides from the world. Any identity can become a persona; that of the scientist is no exception.

The German poet, philosopher, and scientist Johann Wolfgang von Goethe (1749–1832), anticipating Husserl's phenomenological revolution by the better part of a century, provides us with a welcome method for subverting this human, all too human, tendency. He starts with a simple question:

> When contemplating nature, whether in great things or small, I have constantly asked myself the question: is it the object which is here declaring itself, or is it you yourself?

3. See Myron Sharaf, *Fury on Earth: A Biography of Wilhelm Reich* (New York: St. Martin's Press/Marek, 1983), 311–14 and 443.

> And this is also my stance to predecessors and fellow workers.[4]

Much later, philosopher Jean-Luc Marion would call the tendency to find oneself and aspirations in phenomena by another name: idolatry.[5]

To avoid this methodological, political, and, ultimately, spiritual trap, Goethe advocates what he calls "a delicate form of empiricism," a method of inquiry "which enters into the closest union with its object and is therefore transformed into an actual theory," although he cautions us that "this heightening of spiritual capacity belongs to a highly civilized epoch."[6] As David Seamon has observed, "Goethe's way of science was highly unusual because it moved away from a quantitative, analytic approach to the natural world and emphasized, instead, an intimate firsthand encounter between the student and the thing studied."[7] The reciprocity of such a method suggests the reality of what Alfred North Whitehead called "mutual immanence."[8] Goethe's hoped-for epoch has, almost two hundred years later, failed to arrive, and his method is antithetical, heretical even, to

4. Maxim 593, from Johann Wolfgang von Goethe, *Maxims and Reflections*, trans. Elizabeth Stopp, ed. Peter Hutchinson (London: Penguin Books, 1998), 79.

5. Particularly in *The Idol and Distance: Five Studies*, trans. Thomas A. Carlson (New York: Fordham University Press, 2001) and *God without Being: Hors-Texte*, trans. Thomas A. Carlson (Chicago: University of Chicago Press, 1991).

6. Maxim 565. Goethe, *Maxims and Reflections*, 75.

7. David Seamon, "Goethe's Way of Science as a Phenomenology of Nature," *Janus Head* 8, no. 1 (2005): 86–101, at 86–87.

8. Alfred North Whitehead, *Adventures in Ideas* (New York: Macmillan, 1925), 168–70.

what he disparaged as the "gloomy empirical-mechanical-dogmatic torture chamber" employed by the descendants of Francis Bacon and, in particular, of Isaac Newton: a method that has persisted into our own epoch.[9] Indeed, our epoch has proved the apotheosis of the torture chamber.

A hodgepodge of Catholic popular historians of science are fond of pointing to the ways in which Catholic scientists—Gregor Mendel or Georges Lemaitre, for example—have contributed to the success of the Enlightenment paradigm of the hard sciences, as if their contributions constitute some kind of cultural legitimacy. Let me say this clearly: such apologetic moves are the machinations of the desperate. Indeed, these unconvincing apologetic gestures more accurately point to a species of cultural schizophrenia as they reify the secular/religious binary. Some of these same Catholic scholars in their desperate zeal for legitimacy even point to René Descartes and Marin Mersenne as exemplary Catholic scientists. *Pourquoi?* The mere fact that a scientist was a practicing Catholic—usually simultaneously conforming to the norms of his discipline on the one hand and the dogmas of his Church on the other—says nothing about a *Catholic* science. Nothing.

My claim is that Goethe's method is much closer to the notion of a Catholic science than any other method of scientific inquiry presently available. Goethe's delicate empiricism, first of all, is characterized by *Ehrfurcht*, reverence: reverence, above all, for the phenomenon before one, be it a flower, an owl, a geode, the rising sun, or a piece of writing. That is to say, reverence for creation. This is not to suggest that contemporary scientists (or even Goethe's contempor-

9. Maxim 430. Goethe, *Maxims and Reflections*, 55.

aries) do not at some point come to reverence for their subject. I am sure it is quite common. But how different would science be—and our experience of the sciences—if reverence were the foundation for its method? It is not something just for children.

Goethe's method allows the phenomenon to be present to the observer—but also allows the observer to be present to the phenomenon. For Goethe, the so-called "scientific method," with its reliance on experiment and the use of apparatus, rather than allowing phenomena to be present, actually works as a kind of filter or screen between phenomenon and observer—to the point where the phenomenon is in danger of not really being seen or experienced. The reason for this apprehension is that Goethe sees apparatus as focused only on a part of the phenomenon at the expense of the whole:

> Nothing happens in living nature that does not bear some relation to the whole. The empirical evidence may seem quite isolated, we may view our experiments as mere isolated facts, but this is not to say that they are, in fact, isolated.[10]

Living is the key term here: through reverence, the *life* of the phenomenon becomes apparent to the observer. This leads to an observation of Goethe's still applicable to the academy: "In general the sciences put some distance between themselves and life, and make their way back to it only by a roundabout path."[11] Goethe's statement, alas, applies just as

10. Johann Wolfgang von Goethe, "The Experiment as Mediator between Object and Subject (1792)," in *Scientific Studies*, ed. and trans. Douglas Miller, *Goethe's Collected Works*, vol. 12 (New York: Suhrkamp Publishers, 1988), 15.

11. Goethe, *Scientific Studies*, 306.

much to the humanities and social sciences as it does to mathematics and the hard sciences. Nevertheless, attention to life and to the splendor that sometimes shines through phenomena is good Catholic science.

Oddly, science—as well as philosophy—does not quite seem to know what *life* is, or at least it seems to be limited in its understanding of what it terms "life." Evolutionary biologists, for instance, often point to Haeckel's concept of an *Urschleim*[12] (often translated as "primordial soup," but more accurately "primordial slime") as the source of all life on earth and suggest that the experiments undertaken by Stanley Miller and Harold Urey in the early 1950s, in which they bombarded ammonia and methane with energy in order to generate life, indicate some proof of the notion. Some days after their bombardment (note the metaphor), some amino acids and nucleotides appeared in their alembic, though not exactly "living" cells. Nevertheless, evolutionary textbooks often add a magisterial coda to their exegesis of the primordial soup: "Thus life began."[13] There is a better name for theories like this: creation myth.[14]

Coming to terms with what is denoted by the word "life" offers an important epistemological and ontological field of activity in which science, philosophy, theology, and the arts

12. See the discussion on Haeckel's proposal in Mario A. Di Gregorio, *From Here to Eternity: Ernst Haeckel and Scientific Faith* (Göttingen: Vandenhoeck & Ruprecht, 2005), 434–35.

13. Mary Maxwell, *Human Evolution: A Philosophical Anthropology* (New York: Columbia University Press, 1984), 18.

14. Stuart A. Kauffman argues that scientists will be able "to create life anew sometime in the next hundred years," but this is rather the eschatological equivalent of the primordial soup creation myth. See his *Reinventing the Sacred: A New View of Science, Reason, and Religion* (New York: Basic Books, 2008), 46.

can equally participate. That is to say that a multivalent, holographic investigation into the notion of "life" can prove to be the starting point of a truly Catholic science, a truly Catholic philosophy and theology, a true Catholic art.

Into the early modern period, some scientists were still mindful of the holographic nature of the term, but they quickly fell into the minority as the Scientific Revolution advanced, and the Enlightenment eventually eclipsed such a notion almost completely. In the early seventeenth century, for example, the English Catholic natural philosopher and virtuoso Sir Kenelm Digby undertook a vigorous program of research into palingenesis, the laboratory-based attempt at resurrecting a plant or animal from its own ashes. This was not so much a foray into magic as an attempt to discover the biological and chemical qualia of the Resurrection Body. And he claimed to have succeeded.[15] Likewise, the English physician, theologian, and natural philosopher Robert Fludd speculated much on the two understandings of *light* as it appears in the Genesis creation story: one light created on the first day ("Let there be light") and another created on the fourth with the creation of the sun, moon, and stars. Fludd reads the first day's light in the context of the uncreated light, of Christ as the light of the world, and of the Johannine assertion that "In him was life, and the life was the light of men" (John 1:4), of which the created light of the fourth day is but a signifier: "therefor that Light

15. See Kenelm Digby, *A Discourse concerning the Vegetation of Plants, spoken by Sir Kenelme Digby, at Gresham College, on 23. of January, 1660. At a Meeting of the Society for promoting Philosophical Knowledge by Experiments* (London, 1661), 76–78 and 83–88. See also the chapter entitled "Love's Alchemist: Palingenesis and the Unconscious Metalepsis of Sir Kenelm Digby" in Michael Martin, *Literature and the Encounter with God in Post-Reformation England* (Farnham, UK: Ashgate, 2014).

which is in the Sun, doth participate with divinity."[16] For Fludd, "life" and "light" are nearly interchangeable terms.

More recently, Michel Henry has visited this idea in the context of philosophy: "Thus, in its absolute generation, Life generates within itself He whose birth is the self-accomplishment of this Life—its self-accomplishment in the form of its self-revelation."[17] Or, speaking in less theologically informed terms and in a critique of Martin Heidegger: "Phenomenology's great realization is the idea—which comes from Husserl—that the world is in no way limited to the existing world and that basically there is the constant possibility for the installation of a new ontological dimension. Reality is thus not reduced to things, instead there are unsuspected dimensions of being and a characteristic of humankind is to live in these new fields."[18] Materialistic science, we can infer, hamstrings itself—indeed, does itself a terrible violence—by ignoring this very fundamental phenomenological insight. For there is surely something that illuminates the soul, that is the livingness of life, that is antecedent to both the physical light we behold with the senses and the biological life expressed by creatures. And surely it would be to the benefit of contemporary scientists to be attentive to phenomena that are so obvious, and not simply dismiss them under the pretense of calling them aspects of an obsolete vitalism.

But such is obviously not a motivating factor in the cur-

16. Robert Fludd, *Mosaicall Philosophy, Grounded upon the Essential Truth or Eternal Sapience* (London: 1659), 62.

17. Michel Henry, *I am the Truth: Toward a Philosophy of Christianity*, trans. Susan Emanuel (Stanford: Stanford University Press, 2003), 57.

18. Michel Henry, "Art and the Phenomenology of Life (from an interview with Jean-Marie Brohm published in *Prétentaine* December 1996)," trans. Michael Tweed.

rent climate of scientific research. Nor is the modern scientific milieu particularly mindful of the ways in which "the science of one age becomes the mythology of the next"[19] and that the egregore of the status quo seeks preservation from slipping into irrelevance in the face of data incongruous with the wishes of the academic archons. Humility in the face of the unknown and less condescension toward the past could serve science well. It may be, indeed, that the ancient Greeks and the New Testament provide modern science with more apt and flexible vocabularies when parsing the holographic nature of the term "life."

In classical and koine Greek, the terms typically rendered into English as "life" are *bios* (βιος), *zoë* (ζωή), and, to a lesser degree, *psuche* (φυχή). *Bios* in its Greek origins denotes a "way of life," be it the natural inclinations of plants and animals to grow and propagate in their unique and characteristic ways or the manners in which individuals might go about making a *living*; it is a property particular to the time between birth and death.[20] *Zoë*, on the other hand, "is 'non-death,' . . . something that does not even let death approach it . . . [the] 'time of being.'"[21]

In the *Politics*,[22] Aristotle discusses human life in terms of *bios*, a fact to which both Hannah Arendt and Giorgio Agamben have devoted considerable attention, exemplifying our culture's preoccupation with ways of living to the

19. Lynn Margulis and Dorion Sagan, *What Is Life?* (New York: Simon & Schuster, 1995), 18.

20. Hannah Arendt, *The Human Condition* (Chicago: University of Chicago Press, 1958), 12 and 97.

21. Carl Kerényi, *Dionysos: Archetypal Image of Indestructible Life*, trans. Ralph Manheim, Bollingen Series LXV, 2 (Princeton: Princeton University Press, 1976), xxxiv.

22. 1254a.

neglect of *zoë*—which Arendt categorically dismisses as "mere *zoë*."[23] In the New Testament and in Plotinus, however, the emphasis is placed on *zoë* as the integral essence of life. For Plotinus, *zoë* is the source of inexhaustible energy, a quality he associates with a proximity to wisdom:

> Life holds no weariness for anyone when it is pure: and how should that which leads the best life grow weary? This life is wisdom, wisdom not acquired by reasonings, because it was always all present, without any failing which would make it need to be searched for; but it is the first, not derived from any other wisdom; the very being of Intellect is wisdom: it does not exist first and then become wise.[24]

Zoë in Plotinus bears a deep affinity for what later times would call sophiology. But more of this anon.

In the New Testament, the phrase translated into English as "eternal life" is not "eternal *bios*." Rather, it is ζωὴν αἰώνιον, "eternal *zoë*." Likewise, in Christ's proclamation "I am the way, and the truth, and the life" (John 14:6), *zoë* is the life of his identity. By contrast, when Christ says, "No one has greater love than this, to lay down one's life for his friends" (John 15:13), the word translated as life is not *zoë*, nor is it *bios*, but rather *psuche*, an occurrence found also in John 10:11: "The good shepherd lays down his life for his sheep." Both are indications of the life imparted to the body by the soul. St. Paul also speaks of giving up his own life in

23. Arendt, *The Human Condition*, 97.

24. Plotinus, *Ennead* 5.8.55. From Plotinus, *Enneads*, trans. A.H. Armstrong, 7 vols. (Cambridge: Harvard University Press / London: William Heinemann, 1984), 5: 253–54. There is much resonance here with Kerényi's assertion that *zoë* is "indestructible life," a notion absolutely foreign to most contemporary notions of "life" found in philosophy and biology.

terms of *psuche* (Romans 16:4). Only in 1 Timothy 2:2 does Paul use *bios*, and that when describing how one might "lead a quiet and peaceable life in all godliness and dignity." *Zoë*, unlike *bios* or *psuche*, cannot be surrendered, since it is not a particular property of this or that individual person: it is a property of Christ and He imparts it to creation.

The gatekeepers of modern science, not surprisingly, categorically reject *zoë* as life, slandering such a notion under charges of "vitalism" if not accusations of phantasy. And raising the possibility of *zoë*—which smacks so much of mysticism and metaphysics for the scientific materialist—can render an incautious scientist *persona non grata* in the greater scientific community. As Rupert Sheldrake has pointed out, "Scientists know that the doctrines of materialism are the rules of the game during working hours,"[25] whatever their philosophical or religious commitments. Is it not surprising, then, how limited (and, one might say, impoverished) modern scientific and philosophical discourses have been rendered by attending only to *bios* and ignoring *zoë* and *psuche* almost entirely? They claim these do not exist, and so instead attend to what is perishable. We live in a discourse of death.

Hans Jonas was acutely aware of the philosophical, scientific, and cultural problems attendant to the assumption that the notion of "life" has simply been reduced to a utilitarian term fit for experimental application. In Jonas's estimation, science as we now know it (and as we have known it since at least the Renaissance) cannot contend with the mystery of life and so contents itself with the "knowability" of death: "The lifeless has become the knowable par excel-

25. Rupert Sheldrake, *The Science Delusion: Freeing the Spirit of Enquiry* (London: Coronet, 2012), 24.

lence and is for that reason also considered the true and only foundation of reality. . . . Accordingly, it is the existence of life within a mechanical universe which now calls for an explanation, and explanation has to be in terms of the lifeless."[26] Jonas's sentiments are nothing new. Indeed, they have roots deep in the Romantic resistance to the totalizing Enlightenment epistemologies to which we are still inured. Goethe, obviously, was at the vanguard of such resistance, and his delicate empiricism has resonances with the aesthetics of William Wordsworth, who in 1798 spoke with the voice of a prophet to his age as well as to our own:

The Tables Turned

Up! up! my Friend, and quit your books;
Or surely you'll grow double:
Up! up! my Friend, and clear your looks;
Why all this toil and trouble?

The sun, above the mountain's head,
A freshening lustre mellow
Through all the long green fields has spread,
His first sweet evening yellow.

Books! 'tis a dull and endless strife:
Come, hear the woodland linnet,
How sweet his music! on my life,
There's more of wisdom in it.

And hark! how blithe the throstle sings!
He, too, is no mean preacher:
Come forth into the light of things,
Let Nature be your Teacher.

26. Hans Jonas, *The Phenomenon of Life: Toward a Philosophical Biology* (New York: Delta, 1968), 10.

She has a world of ready wealth,
Our minds and hearts to bless—
Spontaneous wisdom breathed by health,
Truth breathed by cheerfulness.

One impulse from a vernal wood
May teach you more of man,
Of moral evil and of good,
Than all the sages can.

Sweet is the lore which Nature brings;
Our meddling intellect
Mis-shapes the beauteous forms of things:—
We murder to dissect.

Enough of Science and of Art;
Close up those barren leaves;
Come forth, and bring with you a heart
That watches and receives.[27]

What Wordsworth calls for here is an honest view of phenomena uncompromised by the intermediaries of intellectualization, the bondages of obligation, and the interference of experimental apparatus. That is, Wordsworth wants us to first attend to what is real, what is before us. In our own time, how much of our experience of the Real is filtered through the technological or the conceptual? Do we even know what the Real is? It is not, I fear, a subject explored in conventional modes of education. Education does not attend to *zoë*. Nor is it concerned with *psuche*. Nor does it attend in any significant way to a living experience of *bios*. We murder to dissect.

27. From *The Poetical Works of William Wordsworth: Evening Voluntaries...*, ed. E. de Selincourt and Helen Darbshire (Oxford: Clarendon Press, 1947).

Rudolf Steiner, deeply influenced by Goethe's scientific writings and remarkably attentive to the phenomenology of things,[28] describes a phenomenon that many of us take for granted: our immersion in the world of electricity, a world of the technological. "You see," says Steiner,

> when we speak of electricity, we enter a sphere that presents a different aspect to the imaginative vision than that of the other spheres of Nature. So long as man remained within the light, within the world of sound, that is to say, in the spheres of optics and acoustics, it was not necessary to judge morally that which appeared in a stone, a plant, or an animal, either as colours in the sphere of light, or as sound in the world of tones; it was not necessary to judge these things morally, because he still possessed an echo, weak though it was, of the reality of concepts and ideas. Electricity, however, drove out this echo. And if today we are, on the one hand, unable to discover a reality in the world of moral impulses, we are, on the other hand, even less able to discover a moral essence in that sphere which is now considered to be the most important constituent of Nature.[29]

Does such an idea ever enter into our consciousness? Do we really believe the "musical" or "visual" phenomena we experience mediated (note the metaphor) through the technological are morally neutral? To my knowledge, contemporary science has not confronted this problem to any significant degree, and, if the current trajectory is any indication, it won't be doing so any time soon. As Sheldrake

28. In the early 1880s, as a young scholar, Steiner edited the scientific writings for the Kürschner edition of Goethe's collected works.

29. "Concerning Electricity: A Lecture by Rudolf Steiner, Dornach, January 28, 1923, GA 220," http://wn.rsarchive.org. Originally published in English in *Anthroposophic New Sheet*, no. 23/24 (9 June 1940).

argues, "Governments and corporations do not usually pay scientists to do research because they want innocent knowledge, like that of Adam before the Fall. Naming animals, as in classifying endangered species of beetles in tropical rainforests, is a low priority. Most funding is a response to Bacon's persuasive slogan 'knowledge is power.'"[30]

My point, however, is not that we should avoid using electricity or technology. My point is only to suggest that *how* we observe phenomena possesses ontological, epistemological, and even physiological ramifications of which we should be vigilantly mindful. Observation, as Werner Heisenberg and Niels Bohr were among the first to acknowledge, is never scientifically neutral. And I would add that neither can it ever be morally neutral.

Our sciences, though, are rather enthralled to the *daemonia* of abstraction, dogmatics, and the faerie glamor of electronic and technological mediation. Recently, for example, I asked the science division at my own small liberal arts Catholic college if I might borrow a few prisms for a demonstration on light and color in a theology course. The science faculty didn't know if the college had any. They certainly couldn't find any. I am sure that students *learn* about optics in their physics classes. Or, rather, I am sure they are *told* about or *read* about optics in their physics classes. Perhaps they see a video on optics. But I am not at all convinced that they *experience* optics in their physics classes. My tale, admittedly, is anecdotal. But it is not, I fear, unique in undergraduate science education.

In his prophetic (and wickedly, wickedly satirical) novel *Brave New World*, Aldous Huxley describes a civilization so enthralled by the technological and so insulated from mean-

30. Sheldrake, *The Science Delusion*, 15–16.

ing and metaphysics that such enthrallment becomes a societal norm: "Nowadays the Controllers won't approve of any new game unless it can be shown that it requires at least as much apparatus as the most complicated of existing games."[31] Think about that the next time you see a news story on the hundreds of people waiting in line (among thousands of such lines) to get the next iPhone.

The enthrallment of the sciences to the technological and their immunity (or so it would seem) from ultimate meaning and metaphysics clearly have something to do with federal and corporate STEM dollars flooding into research departments. The lure of profit and power. But it has even more to do, I think, with their estrangement from philosophy and theology. As such, then, science puts itself in danger of becoming a technological enterprise rather than a human one, an enterprise of power devoid of ethics, utilitarian and as cold and inhuman as the emptiness of outer space, as warm and reassuring as the voice of a robocall. Fear is one of its tools. Entertainment another. Efficiency, yet another. Love never figures into the equation. What would happen if it did?

SOPHIOLOGY

The problem of the so-called "two cultures" is not a conflict between two epistemologies as much as it is one between competing cosmologies. The cosmology of scientific materialism starts with the Big Bang and proceeds to atrophy. Everything in between lies open to investigation and interpretation, though only within the proscriptions first outlined by Bacon and Descartes—that such phenomena and

31. Aldous Huxley, *Brave New World* (1932; repr., New York: Harper Perennial Library, 1998), 31.

their effects are testable, verifiable, reproducible, and predictive: a thoroughly verificationist paradigm.[32] Love, for instance, though it has a place in religious cosmologies, has no role in this one. Should it? Or, rather, shouldn't it?

The mainstream of the scientific community, at least as it now exists, would say no. This is a problem. At best, evolutionary biology, evolutionary psychology, or neuroscience would describe love in terms of the biological imperative, partner selection, courtship displays, chemical reactions—but this is simply a "me Tarzan, you Jane" understanding of love. It lies far away from religious understandings of *agape*, or even of Guillaume Apollinaire's pronouncement that "*L'amour qui emplit ainsi que la lumière/Tout le solide espace entre les étoiles et les planètes.*"[33] But my claim is that love must be taken seriously as not only a metaphysical or psychological category, but, indeed, as a scientific one. As Hans Urs von Balthasar has noted:

> Wherever the relationship between nature and grace is torn asunder in the sense of the . . . dialectical opposition between "knowledge" and "faith," worldly being will necessarily fall under the sign of the constant dominion of "knowledge" and thus science, technology, and cybernetics will overpower and suffocate the forces of love within the world. The result will be a world without women, without children, without reverence for the form of love in poverty and humility, a world in which everything is viewed solely in terms of power or profit-margin, in which

32. Thomas Nagel, *Mind and Cosmos: Why the Materialist Neo-Darwinian Conception of Nature Is Almost Certainly False* (Oxford: Oxford University Press, 2012), 38.

33. "Love fills like light all solid space between the stars and the planets." From "Poème lu au mariage d'André Salmon," lines 40–41.

> everything that is disinterested and gratuitous and useless is despised, persecuted, and wiped out, and even art is forced to wear the mask and features of technique.
>
> But if we view creation with the eyes of love, then we will understand it, despite all the evidence that seems to point to the absence of love in the world. We will understand the ultimate purpose of creation: not only the purpose of its essence, which we seem to make some sense of through the various intelligible relationships among individual natures, but the purpose of existence in general, for which no philosophy can otherwise find a sufficient reason. Why in fact is there something rather than nothing? . . . Only a philosophy of freedom and love can account for our existence, though not unless it also interprets the essence of finite being in terms of love. In terms of love—and not, in the final analysis, in terms of consciousness, or spirit, or knowledge, or power, or desire, or usefulness. Rather, all of these must be seen as ways toward and presuppositions for the single fulfilling act that comes to light in a superabundant way in the sign of God.[34]

Even though quantum mechanics has acknowledged the influence of intentionality in changing the behavior of phenomena, the role of our consciousness in the *actus* of observation is generally de-emphasized in the materialistic scientific literature, clearly an act of willed blindness. As David Bohm and Basil Hiley have proposed, "It is no longer appropriate, in measurements to a quantum level of accuracy, to say that we are simply 'measuring' an intrinsic property of the observed system. Rather what actually happens is that the process of interaction reveals a property involving the whole context in an inseparable way. Indeed, it may be

34. Hans Urs von Balthasar, *Love Alone: The Way of Revelation* (London: Sheed & Ward, 1968), 114–15.

said that the measuring apparatus and that which is observed *participate* irreducibly in each other."[35] Yet the willful blindness persists. If simple attendance to phenomena effects change in their behavior, what would our sympathy or antipathy (love or hate) provoke in the same phenomena? We know from our own experience that our attendance to some phenomena (pornography is a good example) can become psychologically or physically (as well as spiritually) damaging to us; and we likewise know from our experience that attendance to other kinds of phenomena (say Mozart's *Requiem* or a glorious sunrise) can have beneficial effects. Is it presumptuous to think that the inverse does not also happen? Disposition governs outcomes. Can it not be, then, that Goethe's *Ehrfurcht* allows phenomena to reveal themselves in their highest being while simultaneously drawing forth our own highest being in the *actus* of beholding, an *actus* more and more in consonance with love as methodology? (An insight that certainly suggests Empedocles was onto something when he characterized Love and Strife as integral to the workings of the cosmos.)[36] As such, phenomena become truly *present* to us. Without a science that acknowledges such a presence (*parousia*), can we even feign surprise that the discipline has become to a significant degree the standing reserve (Heidegger's *Bestand*) of the medical-pharmaceutical-military-industrial complex?

35. David Bohm and Basil Hiley, *The Undivided Universe: An Ontological Interpretation of Quantum Theory* (London: Routledge, 1993), 6. Emphasis in original.

36. "And all through Wrath are split to shapes diverse;/And each through Love draws near and yearns for each./For from these elements hath budded all/That was or is or evermore shall be." From *The Fragments of Empedocles*, trans. William Ellery Leonard (Chicago: Open Court Publishing Company, 1908), 25.

Some will charge that my proposal here sounds more like mysticism than science, a charge often levied at phenomenology. But I counter that the materialistic, positivistic philosophy that undergirds most of the sciences today has become an archaic structure ossified by its own assumptions, an existential dead end. As Nobel laureate physicist Brian Josephson has been arguing since at least the 1980s, the science of the schools simply disregards a significant portion of reality, if not ultimate reality itself:

> The question then arises whether some future science may be able to cope with this aspect of reality or whether it will remain forever beyond the scope of science. The general aim of science being to gain as full and accurate a picture of reality as possible, one would expect logically that scientists in general would take a keen interest in such questions, just as they do in topics such as those of the fundamental constitution of matter, or of the mechanisms of life. In practice, however, such questions have been almost entirely split off from scientific consciousness, in some cases as a result of an atheistic point of view and in others because present science seems to be so far removed from anything spiritual that the suggestion of the possibility of merging of the two seems improbable or even absurd.[37]

It is becoming increasingly clear that despite (or because of) science's amazing almost daily breakthroughs, the "Big Picture is suddenly too small to make sense."[38]

37. Brian Josephson, "Physics and Spirituality: The Next Grand Unification?" *Physics Education* 22 (1987): 15–19, at 15.

38. Angus Fletcher, *The Topological Imagination: Spheres, Edges, and Islands* (Cambridge: Harvard University Press, 2016), 42.

One way to reclaim a scientific worldview that does not reject the entirety of the cosmos is to welcome poetic perception back into the realm of scientific inquiry. The idea of a poetics of science is nothing new. In the Western tradition, alchemical texts, for one, are rich in poetic language and metaphor. Tibetan medicine, for another, continues to rely on a poetic understanding of the art of healing. For medicine truly is an art. Unfortunately, the past fifty years have seen literary criticism turn more and more into a science (as dreadful a performance as that has been), when it would have been more fruitful to work at turning science into an art. Indeed, could it not be that the gift of poetic perception might salvage science from the instrumentalization of persons and phenomena characteristic of the medical-military-industrial superstructure?

Since at least the time of Francis Bacon, the relationship of the scientist to nature has been driven by an ethos of rape, torture, and pillage. As Lord Verulam writes in *Novum Organum*, "The secrets of nature are better revealed under the vexations of art than when they go their own way."[39] Carolyn Merchant, Mary Midgley, and Pierre Hadot have all examined the philosophical and cultural implications of such an approach, and not flatteringly. For Hadot, this ethos is revealed in our ever-growing fascination with technology and the mechanical disfigurement of nature (and our relationship to nature) that such a fascination facilitates: "Technology allows us to regain the upper hand over nature.... Moreover, mechanics is a technique that consists

39. Francis Bacon, *Novum Organum*, in *Works*, ed. James Spedding, Robert Leslie Ellis, and Douglas Devon Heath, 2nd ed., 14 vols. (London: Longmans, 1868–1901), 4: Book I, Aphorism 98, p. 98.

in tricking nature, by means of instruments fashioned by human beings: machines of all kinds that enable the production of effects apparently contrary to nature."[40]

For Midgley, modern science tried to disguise its inherently violent ethos by expunging the feminine pronoun with which natural science and natural theology had referred to *Natura*, but which has by no means eradicated its violent methodology. She exposes the entire dark project to the clearest sunlight:

> The literature of early modern science is a mine of highly-coloured passages that describe Nature, by no means as a neutral object, but as a seductive but troublesome female, to be unrelentingly pursued, sought out, fought against, chased into her inmost sanctuaries, prevented from escaping, persistently courted, wooed, harried, vexed, tormented, unveiled, unrobed, and "put into question" (i.e. interrogated under torture), forced to confess "all that lay in her most intimate recesses," her "beautiful bosom" must be laid bare, she must be held down and finally "penetrated," "pierced" and "vanquished" (words which constantly recur).[41]

Such a project, she further argues, persists to the current moment.

Finally, Carolyn Merchant concludes that modern science has deprived us of an organic, whole, and holistic view of nature by first viewing it as a female to be subjugated (in contrast to a female to be loved and honored) and then by

40. Pierre Hadot, *The Veil of Isis: An Essay on the History of the Idea of Nature*, trans. Michael Chase (Cambridge: Harvard University Press, 2006) 102–3.

41. Mary Midgley, *Science as Salvation: A Modern Myth and Its Meaning* (London: Routledge, 1992), 77.

interpreting nature as purely "dead" matter (*Mater*), nonliving and therefore fit for exploitation and optimization. As a result, nature, and with it the human person which is a part of nature, is now understood as machine:

> As the unifying model for science and society, the machine has permeated and reconstructed human consciousness so totally that today we scarcely question its validity. Nature, society, and the human body are composed of interchangeable atomized parts that can be repaired or replaced from outside. The "technological fix" mends an ecological malfunction, new human beings replace the old to maintain the smooth functioning of industry and bureaucracy, and interventionist medicine exchanges a fresh heart for a worn-out, diseased one.[42]

Since Merchant wrote these words in 1993, the continued exploitation of nature has been augmented by what I have called elsewhere the technological colonization of the human person. Human biology, alas, is now understood to have as many (inter)changeable parts as any vehicle rolling down the assembly line. Add to this transhumanism's grotesque caricature of the human being which now looms on the horizon, and it is but a short step to presage the end of our species. As Percy Bysshe Shelley writes in *A Defence of Poetry*, "The cultivation of those sciences which have enlarged the limits of the empire of man over the external world, has, for want of the poetical faculty, proportionally circumscribed those of the internal world; and man, having

42. Carolyn Merchant, "The Death of Nature," in *Environmental Philosophy: From Animal Rights to Radical Ecology*, ed. Michael Zimmerman et al. (Upper Saddle River, NJ: Prentice Hall, 1993), 269–83, at 277.

enslaved the elements, remains himself a slave."[43] Our present condition, despite or because of the scientific advances we hear lauded on every side, indeed resembles a kind of slavery.

There is, however, an antidote to the prevailing scientific-materialist death spiral that is our epoch's ruling epistemology. That antidote is sophiology.

Even though I have written many articles and one entire book on the subject, edited another on it, and composed reams of poetry of a sophiological nature, sophiology is notoriously difficult to encapsulate into an elevator conversation. Even for me. Briefly (very briefly), there are essentially two manifestations of sophiology.

One manifestation arrives as the product of a mystical experience with a figure identified with the Sophia (*Hokmah* in Hebrew) of the Old Testament (particularly in Proverbs 8 and 9).[44] There is also a very real connection between this understanding of Sophia and the Shekinah ("the glory of the Lord") of kabbalistic mysticism: "The sparks of the *Shekinah* are everywhere, scattered among the spheres of metaphysical and physical existence."[45] As the book of Wisdom affirms of Sophia, "She is a breath of the might of God and a pure emanation of the glory of the Almighty; therefore nothing defiled can enter into her" (7:25). Such was the experience of Sophia claimed, for example, by the Anglican mystics John Pordage and Jane Lead and their circle in the

43. From *Shelley's Poetry and Prose*, ed. Donald H. Reiman and Neil Fraistat, 2nd ed. (New York: W.W. Norton & Company, 2002), 530.

44. I say "Old Testament" rather than "Hebrew Bible" because Sophia figures significantly in the books of Sirach and Wisdom, which are not included in the Jewish canon of Scripture.

45. Gershom Scholem, *On the Kabbalah and Its Symbolism*, trans. Ralph Manheim (London: Routledge and Kegan Paul, 1965), 115.

seventeenth century and by the nineteenth-century Russian philosopher Vladimir Solovyov, among others.

The other manifestation resides in the philosophical/theological intuition that the Wisdom of God is integral to our experiences of the Truth, Beauty, and Goodness of the Creation—experiences that become palpable due to the reciprocal disclosure characteristic of contemplative engagement with phenomena.

There is an almost obvious Marian dimension to such disclosure: in the way that the Virgin Mary made God palpable to the flesh in Jesus, so the Virgin Sophia makes God palpable in creation and in created works.[46] In a divine reciprocity, the Virgin Mary, the *Sophia creata*, reveals the full mystery of God's Incarnation; whereas God's Wisdom, the *Sophia increata*, reveals the *parousia*, or presence of God. These experienced phenomena can be aesthetic, scientific, theological, philosophical, or liturgical, and they reveal the supernatural dimensions of Things. Clearly, this engagement intrudes into realms typically ascribed to mysticism. And, indeed, in this regard Sophia occupies a *metaxu* between science, art, and the religious. Art often evinces resonances with mysticism. Why should science be any different?

Sergei Bulgakov, arguably the most important sophiologist of the twentieth century, reads Sophia "as the soul of the world . . . the center of all creation."[47] He further distinguishes how Sophia

46. See the chapter entitled "Jacob Boehme's Sophianic Intuitions" in Michael Martin, *The Submerged Reality: Sophiology and the Turn to a Poetic Metaphysics* (Kettering, OH: Angelico Press, 2015), 39–61.

47. Sergei Bulgakov, *The Philosophy of Economy: The World as Household*, trans. and ed. Catherine Evtuhov (New Haven, CT: Yale University Press, 2000), 150.

> shines in the world as the primordial purity and perfection of the universe, in the charm of a child and the enchantment of a fluttering flower, in the beauty of a starry sky or of a flaming sunrise.... These sophic rays are what attract us to "nature," though in fact such a state is supernatural with respect to the present state of creation.[48]

Bulgakov's Sophia represents not a romantic escape from the sometimes cruel and unpleasant realities of nature—he knows creation is fallen—but he nevertheless recognizes the glory of the Lord (Sophia) coextensive with fallen nature. Attentiveness to the sophianic nature of the world is central not only to religion, science, and art for Bulgakov; it is also crucial to a truly human economy: "The purpose of economic activity is to defend and to spread the seeds of life, to resurrect nature. This is the action of Sophia on the universe in an effort to restore it to being in Truth."[49] Without a sophic awareness, argues Bulgakov, we are left with an impoverished set of tools: "*Ratio*, scientific or theoretical reason, emerges from the ruins of the sophic; it becomes the lantern with which we seek the Logos in the nocturnal darkness."[50] The emphasis on "*ratio*, scientific or theoretical reason" characterizes the sciences—even those practiced by ostensibly Catholic scientists—and also defines the manner in which the sciences are taught in nominally Catholic schools, whether K–12, college, or university. The Logos will not be found with the tools now employed by science. Catholic scientists—if they are, in fact, Catholic scientists and not simply scientists who go to Mass—will require tools proper

48. Ibid., 151.
49. Ibid., 153.
50. Ibid., 152.

to such an investigation in order to truly disclose the glory of God participating in all things, *natura pura* be damned.

For a number of reasons, Sophiology is obviously not a way of perceiving that a majority of the current mainstream scientific community will welcome as valid. There is good news, however: that kind of thinking will eventually die out. A valid, human, and thoroughly Catholic science grounded in sophiology is possible; it is indeed necessary. And Goethe shows the way.

Goethe's reputation as a scientist in the current academic milieu is essentially nonexistent. Still, his influence persists at the margins and his specter haunts quantum physics, which is mindful of the effects of intentionality and the absolute impossibility of impartial observation. Our observations of phenomena *disturb* them, inflict a kind of violence upon them. To avoid this disturbance or violence, Paul Feyerabend recommended turning to pure mathematics in order to remove such contamination from the equation.[51] This further removal, however, this disengagement, simply follows the Cartesian error to its logical conclusion: the binary of mind and body, weighted on the side of mind.

Werner Heisenberg, a great admirer of Goethe, traces the problem of disengaged "observation" (it is still observation) not so much to Descartes as to Newton. He does not mince words:

> This sacrifice of the living and immediate understanding, which has been the basis of scientific progress since Newton, was also the real reason for Goethe's bitter struggle

51. Paul K. Feyerabend, *Physics and Philosophy: Philosophical Papers*, vol. 4, ed. Steffano Gattei and Joseph Agassi (Cambridge: Cambridge University Press, 2016), 33–34.

> against Newton's physical optics and his teachings on colour. It would be superficial to neglect this struggle as unimportant, there is good reason for one of the most eminent men using all his power to combat the achievements of Newton's optics. One can only charge Goethe with a lack of consistency. He should have combated Newton's views but he should have said that the whole of Newton's physics, optics, mechanics and gravitational theory was the work of the devil.[52]

It is a question whether Heisenberg was using "work of the devil" metaphorically. I assume he was, though I tend to side with William Blake's assertion that "We who dwell on Earth can do nothing of ourselves; everything is conducted by Spirits, no less than Digestion or Sleep."[53]

The philosopher and social reformer Rudolf Steiner, on the other hand, definitely ascribed this development to destructive supernatural forces. Steiner's thought is far-reaching—if sometimes bewildering—and to the uninitiated it can seem downright bizarre. Nevertheless, his insights into education (the Waldorf movement), medicine, agriculture (biodynamics), beekeeping, the arts, and other facets of human endeavor have stood the test of time; so he is at least worthy of a hearing when it comes to the role of science in our day and age. And his thoughts in this realm prove surprisingly prescient and useful for the notion of a Catholic science.

52. Werner Heisenberg, *Philosophical Problems of Nuclear Science*, trans. F.C. Hayes (New York: Pantheon, 1952), 36–37.

53. William Blake, *Jerusalem, or The Emanation of the Giant Albion*, plate 3. From *The Complete Poetry and Prose of William Blake*, ed. David Erdman, revised edition (Berkeley: University of California Press, 1982).

Steiner, following Hegel's idea of cultural evolution, detects over the arc of history a movement from what he deemed the dreamlike consciousness of the ancient past to the technologically subservient era in which we now live. He limns his timeline in a trajectory of the incarnations of three spiritual beings: Lucifer, Christ, and Ahriman:

> To gain a right conception of the historical evolution of mankind over approximately 6000 years, one must grasp that at the one pole stands an incarnation of Lucifer, in the center the incarnation of Christ, and at the other pole the incarnation of Ahriman. Lucifer is the power that stirs up all fanatical, all falsely mystical forces in human beings, all that physiologically tends to bring the blood into disorder and so lift man above and outside himself. Ahriman is the power that makes people dry, prosaic, philistine—that ossifies them and brings them [into] the superstition of materialism. And the true nature and being of man is essentially the effort to hold the balance between the powers of Lucifer and Ahriman; the Christ impulse helps contemporary humanity to establish this equilibrium.[54]

Clearly, the scaffolding for this proposal echoes Hegelian dialectic. I must admit, though, that I do not always know how to take Steiner's insights—are they imaginations or literal (if spiritual) truths? I do know, however, that his ideas are often very useful, particularly when it comes to describing the luciferic and ahrimanic tendencies in human life, culture, and history.

For Steiner, Ahriman and his deception have two conse-

54. Rudolf Steiner, *The Incarnation of Ahriman: The Embodiment of Evil on Earth: Seven Lectures*, trans. Matthew Barton (Forest Row, UK: Rudolf Steiner Press, 2006), 17–18.

quences: one impacting science and one impacting culture (really two sides of the same phenomenon). First of all,

> Ahriman has the greatest possible interest in instructing men in mathematics, but *not* in instructing them that mathematical-mechanistic concepts of the universe are merely illusions. He is intensely interested in teaching us the concepts of chemistry, physics, biology and so on, as they are presented today in all their remarkable effects, and in making us believe that these are absolute truths, not that they are only points of view, like photographs taken from *one* side. If you photograph a tree from one side, it can be a correct photograph, yet it does not give a picture of the whole tree. If you photograph a tree from one side, it can be a true likeness, yet it does not give a picture of the whole tree as can be gained from photographing it from four sides. Ahriman has the greatest interest in concealing from mankind that in modern intellectual, rationalistic science, in superstitious empiricism, one is dealing with a great illusion, a deception—that men should not recognize this is of the greatest possible interest to Ahriman. It would be a triumphant experience for him if the scientific superstition which infiltrates all areas of life today and which human beings even try to use as a template for the social sciences should prevail into the third millennium. He would have the greatest success if he could then arrive in western civilization in human form and find the scientific superstition as prevailing dogma.[55]

It is no secret that science in our time has, for some, become a belief system, and people with very limited scientific knowledge proclaim that they "believe in science." And even

55. Ibid., 22.

though the corporate investments in scientific "research" have been shown to often compromise not only science but also the health and well-being of millions[56]—to a degree that should certainly undermine the absolute trust some place in the scientific enterprise—our culture persists in its valorization of what "the research shows." This is superstition writ large.

Steiner also describes the cultural ramifications of Ahriman's influence:

> The second means that he employs is to stir up all the emotions that fragment people into small groups—groups that attack one another. You need only look at all the conflicting parties that exist today, and if you are unprejudiced you will recognize that the explanation is not to be found in human nature alone. If people honestly try to explain this so-called universal warfare[57] through human disharmonies, they will realize that it cannot in fact be attributed to physical humanity. It is precisely here that "super-sensible" powers, ahrimanic powers, have been at work.[58]

A cursory visit to an Internet news site, Twitter feed, or Facebook page surely supports such an assertion. Catholics may have other names by which to call what Steiner here designates as Ahriman, but does not such a phenomenon—

56. See, for example, Camila Domonoske, "50 Years Ago, Sugar Industry Quietly Paid Scientists to Point Blame at Fat," *The Two-Way*, https://www.npr.org/sections/thetwo-way/2016/09/13/493739074/50-years-ago-sugar-industry-quietly-paid-scientists-to-point-blame-at-fat, 13 September 2016, accessed 5 June 2018; Tom Felden, "Most Scientists 'Can't Replicate Studies by Their Peers,'" *BBC*, http://www.bbc.com/news/science-environment-39054778, 22 February 2017, accessed 6 June 2018.

57. Steiner delivered this lecture in 1919.

58. Steiner, *Incarnation of Ahriman*, 23.

and the acrimony and fundamentalism, even violence and hate, that accompany it—suggest supernatural influence? Was St. Paul embellishing when he wrote that "our wrestling is not against flesh and blood; but against principalities and powers, against the rulers of the world of this darkness, against the spirits of wickedness in the high places"?[59]

What Steiner says about the gradual incarnation of Ahriman, though couched in exceedingly different terms, is not essentially different from warnings about the surrender of the human soul to the technological found in figures as diverse as Martin Heidegger, Jacques Ellul, George Grant, or, more recently, Giorgio Griziotti.[60]

Finally, Steiner offers us a challenge to find an authentically Catholic science (a science that serves true discovery rather than profit), a science of *what is*, paradoxically pointing to the example of John Henry Newman. Newman, even in 1879, saw the role of Christianity, let alone the Church of Rome, waning in Western civilization, and he hoped for an intervention from on high:

> Hitherto the civil Power has been Christian. Even in countries separated from the Church, as in my own, the *dictum* was in force, when I was young, that: "Christianity was the

59. Ephesians 6:12.

60. See Martin Heidegger, "The Question Concerning Technology," in *The Question Concerning Technology and Other Essays*, by Martin Heidegger, trans. William Lovitt (New York: Harper Torchbooks, 1977), 3–35; Jacques Ellul, *The Technological Bluff*, trans. Geoffrey W. Bromiley (Grand Rapids, MI: William B. Eerdmans Publishing Company, 1990); George Grant, *Technology and Empire* (Concord, ON: Anansi, 1969); Jonathan Crary, *24/7: Late Capitalism and the Ends of Sleep* (London: Verso, 2014); and Giorgio Griziotti, *Neurocapitalismo: Mediazioni tecnologiche e linee di fuga* (Milan: Mimesis, 2016).

> law of the land." Now, everywhere that goodly framework of society which is the creation of Christianity is throwing off Christianity. The dictum to which I have referred, with a hundred others which followed upon it, is gone, or is going everywhere; and, by the end of the century, unless the Almighty interferes, it will be *forgotten*.[61]

How deeply Newman's sense of urgency is echoed in cries of Catholic cultural leaders from our own time who in their existential despair see revelation and divine intervention (and, when they're honest, the hopes for divine retribution on a neglectful and rebellious people) as the only path forward. But this strategy is just as fruitless in our age as it was in Newman's. Even more fruitless.

> Think, for example, of how Cardinal Newman—who, after all, was an enlightened man, one who followed modern religious development—openly stated at the time of his investiture as Cardinal in Rome that a new revelation was needed if the Christian Catholic teaching was to survive. We have no need, however, of a new revelation; the time of revelations in the old sense is over. We need a new science, one that is illumined by the Spirit. But we must have the courage for such a new science.[62]

In other words, we don't need a new revelation; we need to do something with the revelation we already have. Whether or not we have the courage for such a new science is indeed the question before us. Overcoming the egregores of the academy and laboratory and what has been described as

61. *Speech of His Eminence Cardinal Newman on the Reception of the* Biglietto *at Cardinal Howard's Palace in Rome on the 12th of May* 1897 (Rome: Libreria Spithöver, 1879), 7–8.

62. Steiner, *Incarnation of Ahriman*, 30–31.

a culture of "intellectual phase locking" will not be easy. Yet, if science is to be reclaimed as a human endeavor—and not simply as the R&D wing of the corporatocracy—this is what must happen. And such a science will and can only be a truly Catholic science.

2

ART AS ESCHATOLOGY

> After death the body survives, under favourable conditions, for a considerable period. And the dead mollusc leaves its shell behind it. So it has been with the corpse of Christendom.
>
> EDWARD INGRAM WATKIN[1]

CHRISTENDOM died a long time ago. The Oxford Movement, the Gothic Revival, and Traditionalism all failed to restore the golden age when Christ was King (a title used only ironically in the Gospels but embraced with fervor during the too-long centuries of Christianity's flirtation with political power). More recently, in its spectacular failure, the "New Evangelization" has proved itself neither new nor evangelization. Instead, it simply appropriated the modes and attitudes of the master culture it hoped to supplant (if not save) and became a caricature of that culture, prone to its polemics and polarizations under the guise of spreading the Gospel. A sort of voluntary demonic possession.

Christians are afraid of the death of Christianity. This is irony at its most sublime. Perhaps also hypocrisy, as it betrays the fact that we do not believe what we believe.

Fear of death is fear of the To-Come, fear of the future.

1. Edward Ingram Watkin, *Catholic Art and Culture* (New York: Sheed & Ward, 1944), 192.

This fear characterizes humanity, of course, and it is nothing for which to feel shame. Terror accompanies the contemplation of eschatology. As Jacques Derrida put it so honestly,

> But the Messiah might also be the one I expect even while I do not want him to come. There is the possibility that my relation to the Messiah is this: I would like him to come, I hope that he will come, that the other will come, as other, for that would be justice, peace, and revolution—because in the concept of messianicity there is revolution—and, at the same time, I am scared. I do not want what I want and I would like the coming of the Messiah to be infinitely postponed, and there is this desire in me.[2]

No one really wants the Messiah to come. Or so it would seem.

This is nowhere more obvious than in the arts, particularly in the allegedly Christian crystallizations of things called art. Not in every case, of course, but it certainly permeates the atmosphere in the way of stale incense or appears as an excrescence akin to the residual alkaline rings that form around the basin of a disused holy water font. Christian versions of the arts, when not busying themselves with adopting the modes and vocabularies of the master culture, all too often engage in an aesthetics of retrieval, as if Michelangelo and Raphael, for example, in the fine art version of a Great Books program, are the answer to modernity's ills. This is even the case with the all-too-habitual invocations of inhabitants of a more recent past—C.S. Lewis, J.R.R. Tolkien, or Flannery O'Connor—and whose acolytes often see

2. John D. Caputo (ed.), *Deconstruction in a Nutshell: A Conversation with Jacques Derrida*, 2nd ed. (New York: Fordham University Press, 1996), 24–25.

fit to adopt the requisite uniforms of beards, pipes, tweeds, if not crutches and peacocks. These tombs are beautiful, brothers and sisters, but they are still tombs.

Any Christian art worthy of being called "Christian" or "art" should be grounded in the future. John the Baptist, therefore, is the archetype of the Christian artist: for he calls the Messiah out of the future. The Baptist, while simultaneously embedded in a tradition, nevertheless lives in the wild, on the outskirts of that tradition, a prophet of "the wild Being."[3] He is of the tradition, but not in it. He does not listen to criticism. He does not solicit advice. He does not fear death. He does not network or build his résumé.

But, as Nikolai Berdyaev reminded us, historically, Christianity has not exactly embraced the act of creativity. Indeed, prohibitions concerning graven images have haunted (and continue to haunt) the inheritors of monotheistic religions, even those cultures now recognized as secular, an anxiety as persistent today as it was during the iconoclast controversy and the Reformation. But whereas postmodernity merely dismisses the notion of transcendental beauty in favor of the ephemeral fashions of attraction, the Reformation exhibited its anxiety through its ambivalent equivalences of beauty with sin, an idea that still troubled Jacques Ellul.[4] But it's

3. A phrase of Maurice Merleau-Ponty. See his *The Visible and the Invisible, followed by Working Notes*, trans. Alphonso Lingis, ed. Claude Lefort (Evanston, IL: Northwestern University Press, 1968), 121.

4. For Ellul, iconoclasm "is the first act of the Christian life." See his *Humiliation of the Word*, trans. Joyce Main Hanks (Grand Rapids, MI: William B. Eerdmans Publishing Company, 1985), 96. See also Stanley Fish's examination of this dialectic in John Milton in *Surprised by Sin: The Reader in* Paradise Lost, 2nd ed. (Cambridge: Harvard University Press, 1998).

more than this. Neither the Gospel nor the Fathers speak of human creativity. Writes Berdyaev, "Creativeness is something mysterious and hidden. The revelation of creativeness does not come from above but rather from below—it is an anthropological, not a theological, revelation."[5]

But it is still a revelation. This was clearly the prophetic space inhabited by William Blake, but such an understanding of the role of creativity need not be the province of antinomian mystical enthusiasm. Human creativity is participation in God: as "it wills another world, it continues the work of creation."[6] As such, true creativity—a creativity touched by the mystery of redemption—becomes theurgy. And it is in the theurgic that art speaks the language of eschatology: "Theurgy is the banner of the art of the last times, the art of the end."[7] In its eschatological vocation, art anticipates the revelation of the mystery of the glorified body.

But, we hasten to admit, this is precisely not what art has become in a culture dominated by both market-driven metrics and Marxist materialism.

When Walter Benjamin wrote in 1935 that "mechanical reproduction of art changes the reaction of the masses toward art," he had no way of anticipating what the intervention of cyberspace would do to art.[8] But he certainly saw the cultural repercussions in the flattening of art as a thing

5. Nicolas Berdyaev, *The Meaning of the Creative Act*, trans. Donald A. Lowrie (New York: Collier Books, 1962), 92.

6. Ibid., 95.

7. Ibid., 239.

8. Walter Benjamin, "The Work of Art in the Age of Mechanical Reproduction," in *Illuminations: Essays and Reflections*, ed. Hannah Arendt, trans. Harry Zohn (New York: Schocken Books, 1968), 234.

for mass consumption. And not only is art threatened: human culture in general—*cultus*—also suffers. Martin Heidegger saw the dangers of the factory and assembly line for the human arts as well as for human flourishing:

> What threatens man in his very nature is the view that technological production puts the world in order, while in fact this ordering is precisely what levels every *ordo*, every rank, down to the uniformity of production, and thus from the outset destroys the realm from which any rank and recognition could possibly arise.[9]

Heidegger and Benjamin could never have imagined the MFA production line that all too often turns the pursuit of the making of art or the making of poems into cliquish schools for dilettantes or political coteries more interested in making "statements" (or money—the two are not mutually exclusive) or the laughable and self-aggrandizing concern to "raise awareness" at the cost of authenticity and ability, so much so that the arts in our age have been identified as under the zeitgeist of a "post-skill movement." In the music "industry" (an apt word), sampling and Auto-Tune have likewise flattened the need for skills or vision in that realm, technological innovation obviating the need for them.

Coupled with these repercussions of the rise of the technological are the likewise pernicious demands that the technological—particularly through the exponential growth of social media—has surreptitiously injected into human persons in general, let alone artists and poets, to turn themselves into products to be consumed. "Gone viral," indeed. To be sure, the temptation to become "a sort of prostitute, a

9. Martin Heidegger, *Poetry, Language, Thought*, trans. Albert Hofstadter (New York: Harper & Row, 1971), 117.

sort of courtesan, a sort of lap dog, a person existing only for the pleasure of his patrons"[10] was always a danger for the artist, but now this danger touches all of us. But at least the prostitute gets paid. On social media and in the proliferation of creative and artistic works available on the Internet (whether pirated or not), those really getting paid are not the creators. "We need more clicks" is the prayer of a demon.

The theurgic capacities of art function in a way exactly opposite to those forces that would turn all art, all artists, and, ultimately, all persons, into commodities. In the theurgic act, according to the Neoplatonist Iamblichus,

> unutterable truths are expressed through secret symbols, beings beyond form brought under the control of form, things superior to all image reproduced through images, and all things brought to completion through one single divine cause, which itself so far transcends passions that reason is not even capable of grasping it.[11]

Indeed, I have often wondered if it is not the fetishization of reason so common among certain postmodern neo-Thomists and Calvinists (among others) that stains revelation with the blood of ego and thereby corrupts the appearance of the very thing they wish to reveal. Art, as it reaches beyond reason and toward mystery, certainly could be a higher form of theology. Unfortunately, it, too, succumbs to the egoistic temptation of rationality (often, again, in the form of making a "statement") and likewise partakes in the hubris particular to those using scientific "consensus" (the

10. Eric Gill, *Beauty Looks After Herself*, introduction by Catherine Pickstock (1933; repr., Ranchos de Taos, NM: Sophia Perennis, 2012), 106.

11. Iamblichus, *De mysteriis*, trans. Emma C. Clarke, John M. Dillon, and Jackson P. Hershbell (Atlanta: Society of Biblical Literature, 2003), 1.21, 79–81.

egregore) as a protection from the very real dangers of authentic engagement.

As makers of images, artists are implicated in works of magic, calling forth spiritual presences into the materials of their making. No doubt, this fact has deeply underscored the anxieties of Reformers and other iconoclasts about the arts. Their anxieties have transmogrified over time (as Reform has transmogrified into secularism) to the point where the magical and theurgic dimensions of art are all but ignored, save for the potential some forms of art might have for propaganda or mass sedation. Nevertheless, as John Milbank has argued, "In coming to know by working, we arrive at new vision through the images we have made, the songs we have sung, the words we have uttered, and this 'seeing through' is the theurgic invocation of the divine by which alone God can descend to us—the infusion of our own works with his inspiration."[12] This theurgic capacity is not a supplement to Christian life, but rather is implicit in the promises of Pentecost.[13]

This is not to say, of course, that all spiritual presences are beneficent. And that truth, to some extent, explains the Reform's "censorship of the imaginary."[14] But it does not excuse it. On the other hand, a completely materialist aesthetic is an impossibility. All art possesses a spiritual correl-

12. John Milbank, "Sophiology and Theurgy: The New Theological Horizon," in *Encounter between Eastern Orthodoxy and Radical Orthodoxy*, ed. Adrian Pabst and Christoph Schneider (Farnham, UK: Ashgate, 2009), 45–85, at 66.

13. Sergius Bulgakov, *Unfading Light: Contemplations and Speculation*, trans. and ed. Thomas Allan Smith (Grand Rapids, MI: William B. Eerdmans Publishing Company, 2012), 284.

14. Ioan P. Couliano, *Eros and Magic in the Renaissance*, trans. Margaret Cook (Chicago: University of Chicago Press, 1987), 222.

ative. The danger lies in pretending it does not. When art ceases to be theurgy, other spiritual realities, not all benign, can usurp its theurgic dimension.

The philosopher, poet, and yogi Sri Aurobindo (1872–1950), educated in England and deeply interested in English poetry, was profoundly attentive to "the point at which the aesthetic, the vital, the intellectual elements of poetic speech pass into the spiritual" and wrote much on the varying degrees to which poets have stumbled upon or even habitually risen to what he called, coming from the Hindu tradition, the power of the Mantra: "Not in its substance or its form a poetic enunciation of philosophic verities, but a rhythmic revelation or intuition arising out of the soul's sight of God and Nature and itself and of the world and of the inner truth—occult to the outward eye—of all that peoples it, the secrets of their life and being."[15]

The Catholic writer and mystic Valentin Tomberg, himself no stranger to Vedantic philosophy, also wrote about the vocation of the poet and the significance of the poetic utterance. Of primary importance for the poet, according to Tomberg, is that he or she be completely *incarnated*:

> This is why it is necessary to be *incarnated*, i.e. to have the pulsation of warm earthly blood, in order to be able to create poetic works—and not only poetic works of subjective significance (*setrams*) but also those of objective significance (*mantrams*). It was necessary to be immersed in warm human blood, i.e. to be incarnated, and to rise above it by uniting with the luminous blood of heaven, i.e. with hope, in order that the Psalms of David, for example, could be born. It was not in heaven, but rather on earth,

15. Sri Aurobindo, *The Future Poetry with "On Quantitative Metre"* (Pondicherry, India: Sri Aurobindo Ashram, 1997), 31 and 36.

> that the Psalms of David came to birth. And once born, they became an arsenal of magical mantrams not only on earth but also in heaven. Because the mantrams, i.e. magical formulae, of the Psalms are in use as such not only amongst beings with warm blood, i.e. human beings, but also amongst beings with luminous blood—entities from the celestial realms.[16]

And if anyone doubts what he says here, just try speaking the first two verses of Psalm 67/68 whenever in the presence of evil.[17] See what happens.

What Aurobindo and Tomberg say of poetry and poets applies equally to all of the arts and their practitioners. And the point about being incarnated could not be more important.

Some artists and poets, almost by caricature, do not quite seem to be, so to speak, all the way in their bodies; and thus their work likewise lacks a requisite congruence with reality. They and their work appear almost diaphanous, disembodied, otherworldly. As a result, the effect of their work is thin and watery, ephemeral, profound of intention but void of substance. Others seem to be too much in their bodies: they and their work suffer from a hardness, an aesthetic sclerosis, a heaviness that draws them too deeply into subnature. On the other hand, the relatively recent phenomena of the Internet and virtual art spaces, while embodying the flattening of the reception of art with which Benjamin was so concerned,

16. [Valentin Tomberg], *Meditations on the Tarot: A Journey into Christian Hermeticism*, trans. Robert Powell (Amity, NY: Amity House, 1985), 477.

17. "Let God arise; let his enemies be scattered. Let them that hate him flee from before his face.

As smoke vanishes, so let them vanish away. As wax melts before the fire, so let the wicked perish at the presence of God."

also disclose an art that is simultaneously unincarnated and overincarnated: it never takes on a proper body characterized by warmth, while at the same time, because of the electronic, digitized, and pixelated nature of its appearance, it is too enmeshed in *materia*. But art that reaches to the Mantra is thoroughly grounded in the warmth of the blood and the things of this world while at the same time remaining permeable to the *ruach hakodesh* (רוּחַ הַקֹּדֶשׁ).[18]

Even artworks predating the rise of the Internet, because they are synthesized and reproduced *ad infinitum* in that cybermedium, are in danger of spiritual atrophy, their splendor exhausted and cheapened by overexposure. "Works of art," writes Hans Urs von Balthasar, "can die as a result of being looked at by too many dull [*geistlos*] eyes, and even the radiance of holiness can, in a way, become blunted when it encounters nothing but hollow indifference."[19] And one can only imagine what effect works of art mass-produced on T-shirts and coffee mugs has had on their radiance. Our adversary is inordinately wily. He draws the light from luminous bodies and leaves them hollow.

So, not only Christendom, but art, too, is dead. This is especially the case with many so-called "conservative" artists, craftsmen, even philosophers, who apparently think that saving their imagined past is a fight worth having. As a result, they regurgitate the forms of the past—medieval, Renaissance, Mannerist, what have you—in an ethos that

18. John Paul II calls the Holy Spirit "the mysterious Artist of the universe." See *Letter of His Holiness John Paul II to Artists* (1999), https://w2.vatican.va/content/john-paul-ii/en/letters/1999/documents/hf_jp-ii_let_23041999_artists.html.

19. Hans Urs von Balthasar, *The Glory of the Lord: A Theological Aesthetics*, vol. 1, *Seeing the Form*, ed. Joseph Fessio and John Riches, trans. Erasmo Leiva-Merikakis (San Francisco: Ignatius, 1982), 23.

also touches education in the "Great Books" project. I am not at all opposed to the great works of the past (I do have a doctorate in early modern literature, after all), but neither am I interested in a "preservative" aesthetic that tries to retrieve a past that never existed in the first place. Such preservative movements, at their best, are like traveling museums. But museums—as beautiful and important as the works they house may be and as valuable as is their vocation to hold the past—are nevertheless graveyards of culture. Conservatives have tried to hold on to an idea of art and literature in the same way they have tried to hold on to an idea of Christendom. The term "culture vulture" really says a lot in this regard. Retrieval movements may attract followers and bestow a sense of community, but they don't solve anything. The point is not in retrieval, but in finding what was lost. And what was lost is always/already present. Let the dead bury their dead.

Nevertheless, art, in its eschatological dimensions, has the capacity to foreshadow the glorified body promised in the resurrection. Unlike the glorified body, however, art's transfiguration, as von Balthasar indicated, is not necessarily permanent. The transfiguration of art, the art that calls the Messiah from the future (there is no sense in calling the Messiah from the past), transfigures the *materia*—wood, stone, clay, paint, sounds, words, movement—into its glorified form.

An eschatological art possesses all the qualities that Aquinas attributes to the Resurrection Body—impassibility, subtlety, agility, radiance—and the wonder and truth it imparts produce palpable changes in the individuals it touches.[20] Sergius Bulgakov's words on the Resurrection

20. *Summa Theologiae*, suppl., 82–85.

Body also serve well as a description of an eschatological art: "Life in the body with its sensuousness is not abolished but becomes innocent and immaculate, a holy life together with nature, which becomes a source of joy and admiration."[21]

During the medieval and early modern periods, alchemists actively pursued a scientific investigation into the Resurrection Body (a project which quantum mechanics may someday renew). And it is no accident that the alchemists thought of themselves as artists. As Sir Thomas Browne (1605–1682) argues in a discussion on palingenesis,

> To a sensible Artist the formes are not perished, but withdrawne into their incombustible part, where they lie secure from the action of that devouring element. This is made good by experience, which can from the ashes of a plant revivifie the plant, and from its cinders recall it into its stalk and leaves againe. What the Art of man can doe in these inferiour pieces, what blasphemy is it to affirme the finger of God cannot doe in these more perfect and sensible structures? This is that mysticall Philosophy, from whence no true Scholler becomes an Atheist, but from the visible effects of nature growes up a reall Divine, and beholds not in a dreame, as *Ezekiel*, but in an ocular and visible object the types of his resurrection.[22]

And as alchemist, swashbuckler, and Catholic apologist Sir Kenelm Digby affirmed, "I can not place the resurrection of our bodies among miracles, but do reckon it the last worke

21. Sergius Bulgakov, *The Bride of the Lamb*, trans. Boris Jakim (Grand Rapids, MI: William B. Eerdmans Publishing Company, 2002), 448.

22. Sir Thomas Browne, *Religio Medici: A New Critical Edition with Biographical and Critical Introduction*, ed. Jean-Jacques Denonain (Cambridge: Cambridge University Press, 1955), 63.

and periode of nature; to the comprehension of which, examples and reason may carry us a great way."[23] To which he adds, "Glory neither destroyeth nor drowneth nature, but refineth her and then leaveth her att liberty to exercise all her orderly functions; of wch love is the noblest."[24] Teilhard de Chardin's notion of the Omega Point, "the consummation of the Parousia," clearly resonates with such a sensibility.[25] Does even today's artist not possess the potential to transfigure the material of his or her making into its own glorified form? In this way art, at the very least, is eschatology. But when art rises to the level of the Mantram, it is this even more.

Unfortunately, this vocation of art has been obscured by scientific materialism, postmodern ennui, and the absolute commodification of art as well as the artist. We need to renounce these demonic temptations and recapture "the virtue of living dangerously."[26] The making of art comes with a risk, as the Reformers and iconoclasts were so aware, but it is a risk worth taking. As Berdyaev writes, possessed of the mantle of a prophet:

> *The third creative revelation in the Spirit will have no holy scripture; it will be no voice from on high; it will be accomplished in man and in humanity—it is an anthropological revelation, an unveiling of the Christology of man.* God awaits the anthropological revelation and man must not expect to have it from God. And one cannot merely wait

23. Vittorio Gabrieli (ed.), "A New Digby Letter-Book: 'In Praise of Venetia,'" *National Library of Wales Journal* 9, no. 2 (1955): 455.

24. Ibid., 147.

25. Pierre Teilhard de Chardin, *The Divine Milieu: An Essay on the Interior Life*, trans. Bernard Wall (New York: Harper & Row, 1960), 136.

26. Berdyaev, *The Meaning of the Creative Act*, 101.

> for the third revelation; man must accomplish it by himself, living in the Spirit; accomplish it by a free, creative act.[27]

Don't wait for someone else to act first.

Many have written about the sacramental character of the arts. What I have in mind here certainly participates in such an idea. But the sacramental is concerned primarily with the present and from the point of view of the receiver, whereas the eschatological attends to the *telos* and its coming into being. This coming into being from the future is the source of the transfigurative power of art, though as it moves into the present its absolute power subsides and it is only able to reveal itself in contingency. The eschaton, the *parousia*, is nevertheless always/already present, an actualization in time of the timeless, the appearance of the sophianic. "Sophia, the true Creation or creation in the Truth," writes Pavel Florensky, "is a preliminary hint at the transfigured, spiritualized world as the manifestation, imperceptible for others, of the heavenly in the earthly."[28]

In the gnostic mythos, Sophia suffers in captivity (in matter, the world) awaiting her release and redemption by the soul awakened in Christ. Our own agapeic attention to creation, to the arts, to liturgy, Scripture, and so forth, likewise awakens Sophia and simultaneously makes present the *parousia* (certainly this is a theme of Blake's masterpiece *Jerusalem*). Otherwise, she remains in captivity. The attention of the artist is the paradigmatic site of this kind of awakening, which is why even works not ostensibly religious

27. Ibid., 101. Berdyaev's emphasis.

28. Pavel Florensky, *The Pillar and Ground of the Truth: An Essay in Orthodox Theodicy in Twelve Letters*, trans. Boris Jakim (Princeton: Princeton University Press, 1997), 283.

can disclose truth and beauty. As is all too obvious, many artists working out of a religious sensibility are often unable to awaken Sophia from her slumber. Perhaps their zeal becomes an obstruction, an idol reflecting the artist back to him- or herself. In Jean-Luc Marion's words, there is a real danger in pursuing this path: "Subsidizing the absence of the divine, the idol makes the divine available, secures it, and in the end distorts it."[29] We are, indeed, living dangerously. Reformers and iconoclasts cut their potential losses by avoiding such a risk altogether. But such is a coward's economy. The artist reaches hero-like into the fire, risking perdition as well as transfiguration.

Art, all art, is, by nature, an engagement in magic. But when art becomes theurgy it becomes something more, "for magic operates only within the bounds of nature, manipulating and exploiting these forces rather than demonstrating the causative power which lies beyond them," which power theurgy reveals.[30] Theurgy, then, is concerned with the miraculous.

It may be that the theurgic act of the artist, the act which touches divinity, like Sophia, sleeps in the artwork until enlivened by the agapeic attention of the receiver, though I am hesitant to place conditions on the movements of the divine: the spirit bloweth where it listeth. It may also be that even the most mundane of phenomena become art, become theurgy, through the descent of the divine (for example, in the appearance of the Virgin in a garbage dump at Lourdes). Reception certainly has an active role here. Nevertheless, art

29. Jean-Luc Marion, *Idol and Distance: Five Studies*, trans. Thomas A. Carlson (New York: Fordham University Press, 2001), 7.

30. Emma C. Clarke, *Iamblichus' "De Mysteriis": A Manifesto of the Miraculous* (Aldershot, UK: Ashgate, 2001), 21.

that rises to theurgy likewise participates in the eschaton: it foreshadows the glorification of all things.

Art—whether the fine arts, poetry, architecture, music, dance, philosophy, or politics—makes the invisible visible, makes the inaudible apparent, and gives substance to the formless. At its core, it is an act of adoration, though often the artist is only vaguely aware of what it is he or she adores. That which is revealed wishes to reveal itself. Every artistic act is a spiritual exercise; each one engages spiritual realities. Some of these realities seek incarnation through artworks for nefarious ends. But the sublime subject of art, that which approaches from the future, bears with it powers of transfiguration. Its arrival comes to us not as an absolute transfiguration, to be sure, but the spark of transfiguration it carries contains enough light to awaken the dead, if only momentarily.

> We are constantly forgetting that the supernatural is a ferment, a soul, and not a complete and finished organism. Its role is to transform "nature"; but it cannot do so apart from the matter which nature provides it with.
>
> PIERRE TEILHARD DE CHARDIN[31]

31. Teilhard de Chardin, *The Divine Milieu*, 136.

3

THE REIMAGINATION OF EDUCATION

AND THE POSTMODERN SOPHIOLOGICAL HEDGE SCHOOL

EDUCATIONAL reform movements are as common as toadstools after a storm. In most cases, they are nearly as ephemeral. Unless, that is, the government gets involved: in which case they are as resilient and as toxic as plutonium. And the government always gets involved. It is no wonder, then, to find Ivan Illich argue that "the escalation of schools is as destructive as the escalation of weapons but less visibly so."[1] When it comes to education, the hand that rocks the cradle is also the hand that signs the declaration of war. This is a pathological state of affairs. And it doesn't need to be this way.

Regardless of governmental interventions in actual curricula, "expectations," "benchmarks," and—how I loathe the term—"best practices" (pernicious in the United States, and increasingly so since the famous—and failed—"No Child Left Behind" disaster and its offspring), there is always an interest in making education more meaningful. Government intervention only serves to fundamentally betray the learning process. We always know something is wrong with

1. Ivan Illich, *Deschooling Society* (New York: Harper & Row, 1971), 14.

whatever system is in place, but there is almost never enough will to stand against the oppressive environment created by government regulation, accrediting agencies, and, even more formidable, cultural assumptions concerning what education actually is. And what exactly education is or what it is for seems to be almost anyone's guess.

Conventional wisdom often suggests that education is purposed to create a citizenry capable of competing in a global marketplace: this commitment betrays the intentions of social engineers, governments, and corporate interests. It may be conventional, but it is in no way representative of wisdom, a fact that becomes uncomfortably obvious as we watch the humanities drop from college and university curricula at an unnerving rate and liberal arts colleges disappear, never to return. Despite our own era's preoccupation with STEM disciplines and our ignorance of the hypernormalization that accompanies it, it is still possible to recapture just what this educational enterprise is in the first place. And for this, Jacques Maritain articulates the purpose of education in a most succinct manner (paraphrasing Pindar): the purpose of education lies in "*becoming who we are*."[2] I regret to report that most educational models fail to do this.

Most educational models fail to do this because those that dominate serve too many masters and serve all of them poorly—and most of the masters aren't worth serving in the first place. Kieran Egan, one of the most important (and neglected) educationalists of the last twenty-five years, describes the pedagogical schizophrenia that troubles nearly

2. Jacques Maritain, *Education at the Crossroads* (1943; repr., New Haven, CT: Yale University Press, 1960), 1. Maritain's emphasis.

every school system in existence.[3] For Egan, what plagues contemporary education are what he calls "the three old ideas": socialization, Plato and the Truth about Reality, and Rousseau and Nature's Guidance. The problem is that the three are at odds with one another. How, for example, could a socialization model, the aims of which are to mold children into citizens reflective of and responsive to a nation's needs, square with the Platonic ideal of creating a freethinking, independent, and self-actualized individual through mastery of content? Rousseau's notion of letting the child's instincts guide learning is likewise at odds with both socialization and the Platonic mastery of content.[4] What Egan proposes instead is a model that recapitulates the development of cognitive modes over history and in child development, a sort of cultural correlative of Haeckel's "ontogeny recapitulates phylogeny." Egan identifies these as "mythic understanding" (seen in children up to the age of ten or so), "romantic understanding" (ten through fifteen), and "philosophic understanding" (a thinking in systems, upper high school through early college age). He also identifies what he calls "somatic understanding" (in preverbal children) and "ironic understanding" (adult), but his focus is on those types generally involved in K–12 education. There is much to recommend his model, which is probably why it has been almost entirely ignored by the mainstream. But "becoming who we are" figures into it only indirectly.

Unfortunately, the Catholic/Christian response to mainstream education partakes of all the absurdities and short circuits of the master culture that Egan lists. I have spent

3. Kieran Egan, *The Educated Mind: How Cognitive Tools Shape Our Understanding* (Chicago: University of Chicago Press, 1997).

4. Ibid., Chapter 1.

most of my life in and around Catholic education, though I did spend a few years of both elementary and high school in public education and received my doctorate from a public university. It really doesn't matter, though, because there is virtually no difference between the two. One has prayers at the beginning of the day and has its inmates attend Mass upon occasion, and sometimes has a crucifix in the room or a statue of the Virgin. But that's it. Though mission statements often attest to institutions being rooted in "Catholic traditions" of service, or social justice, or "the Catholic intellectual tradition," or some other abstraction, comparing the Catholic educational model to its secular counterpart is by and large an exercise in dissecting minutiae (and what are mission statements anyway—outside of requirements of accrediting bodies run by well-meaning if misguided people upon the MBA model?). Indeed, the secular educational institution has become the superego of contemporary Catholic education, preschool through university. It does not have to be this way.

Even the recent "innovation" in Catholic education (homeschool as well as parochial) of turning to a classical model as a way to restore the glory that was Catholicism partakes of the absurdities Egan delineates. Is it a Platonic search for the truth about reality? Within limits, I guess. Is it socialization? Yes (but to a subculture). Is it trusting in the child's innate sense of wonder? Not a chance (except in the case of Sofia Cavalletti's work, which some graft onto a classical curriculum). In general, though, this is a *Let's make Catholicism great again!* approach that appeals to tribal instinct, a bunker mentality, and nostalgia. Please, let it die.

This nostalgia for the classical—the trivium and the quadrivium—even haunts some very interesting and commendable educational philosophies. Maritain, for instance, in *Education at the Crossroads*, introduces a reconfigured trivium and quadrivium (trivium: eloquence, literature and poetry, music and the fine arts; quadrivium: mathematics, physics and natural science, philosophy, and ethics/politics). Likewise, and certainly inspired by Dorothy Sayers, Stratford Caldecott adopts a trivium/quadrivium template in his model.[5] Caldecott, even though he recognizes "the need to recover a 'poetic' way of knowing,"[6] nevertheless, like many another Catholic educational reformer, seems to feel obligated to hold on to the trivium/quadrivium model. I think this occurs by a mistaken assumption that Catholicism is somehow synonymous with the medieval, an assumption prevalent in Roman Catholic (and even some Protestant) thinking and not at all characteristic of Eastern Catholic or Eastern Orthodox thinking. In practice, however, it approaches the point of becoming a sales gimmick. Whatever the case, the trivium and the quadrivium cycle through a never-ending sequence of the eternal return. It is time to let this way of thinking go; it has become a hindrance.

One problem of the classical model, however modified or embellished, is that it privileges the intellect over all other human capacities, the underlying assumption being that education is essentially a system of intellectual data-gathering and manipulation. To paraphrase Flannery O'Connor, "If that's what education is, then to hell with it." Another problem is that the classical model, we need to admit, is the stereotype of a "proper education." As the education of choice for elites over the last several centuries, its reputation

5. See his *Beauty in the Word: Rethinking the Foundations of Education* (Kettering, OH: Angelico Press, 2012).

6. Ibid., 11.

surreptitiously furthered in novels, television, and film, the classical model is embedded in the cultural imaginary as the exemplum of the crowning glory of education. Certainly, it is a fine education, but, as John Ruskin wrote, "The great leading error of modern times is the mistaking erudition for education."[7] But there is no reason to assume a classical education is the high water mark of Western civilization's contribution to learning. Even St. Augustine, himself a reformed professor of rhetoric from a time when a liberal education was truly a liberal education, eventually thought of it as overrated if not a danger to the soul.[8]

Because of their dissatisfaction with received notions about what constitutes a proper form of education, many educational and social reformers have sought to bring a renewed freshness and vitality to the educational project. Not surprisingly, these initiatives range from the sublime to the ridiculous, while some, such as the Summerhill School, are a mixture of both. Indeed, the entire homeschooling phenomenon—whether in the spirit of Charlotte Mason, John Holt, Rousas John Rushdoony, or whomever—was born of a dissatisfaction with the status quo. Nevertheless, the superego of education as generally practiced haunts, and, indeed, inhibits, the possibility of truly changing anything—and homeschooling is no exception—even as reformers set up shadow systems reminiscent of something from the science fiction of Philip K. Dick. Legislatures,

7. John Ruskin, *The Stones of Venice*, vol. 3 (London: George Allen, 1900), 230.

8. See Kevin L. Hughes, "The 'Arts Reputed Liberal': Augustine on the Perils of Liberal Education," in *Augustine and Liberal Education*, ed. Kim Paffenroff and Kevin L. Hughes (Aldershot, UK: Ashgate, 2000), 95–110.

accrediting bodies, schools of education, and teacher unions, meanwhile, dedicate themselves to maintaining things as they are, upholding "the hidden curriculum of schooling" that "initiates the citizen to the myth that bureaucracies guided by scientific knowledge are efficient and benevolent."[9] They talk a good game about educational reform, but treat only superficial and cosmetic symptoms—which bathroom to use, which informational technology to adopt, what kinds of diversity will be allowed in the classroom—while ignoring causes. As indentured as they are to Egan's "the three old ideas" (and in particular to socialization), they are unable and unwilling to really do something to transform education. We shouldn't look to them for leadership because, honestly, they are happy with things as they are. Any person of integrity would do well to entirely ignore them.

Catholic education, I would suggest—once the model of innovation and spiritual renewal—has pretty much stagnated since the Jesuit reforms in education instituted during the early modern period. Catholic education at the primary and secondary levels, for example, is often underscored by apologetics (the almost obligatory pro-life speeches rehearsed in high school, for instance), which is an ill-fitting accessory combined with the fundamentally secular curricula utilized in Catholic schools. It is not a very successful model, at least as far as apologetics is concerned, and the occasional attempts to inject "creation science" into curricula to somehow signal a sort of religious authenticity proves even more of a failure, as ineffective as it is risible. At the college or university level, unfortunately, the project is often

9. Illich, *Deschooling Society*, 106.

aimed at proving how un-Catholic the intellectual climate of the school is as a way to demonstrate legitimacy in the eyes of the secular educational archons. As a result, far too many people come away from Catholic educational settings as agnostics, if not headed in the direction of atheism.

Rudolf Steiner (1861–1925) gave us a way to at least begin reimagining education with his introduction of Waldorf education. Steiner began this project in 1919 following the ravages of World War I, when Emil Molt, owner of the Waldorf-Astoria cigarette factory, came to Steiner and asked him to develop a system of education for the children of his workers. The inhumanity of the war compelled both of them to reimagine a way of education that would create a more humane world mindful of the glory of God. The classical curriculum, despite its strengths and its permeation of European pedagogy, was unable to create a society of goodness. As Wilfred Owen (1893–1918), himself the recipient of an exemplary classical education, wrote,

> If you could hear, at every jolt, the blood
> Come gargling from the froth-corrupted lungs,
> Obscene as cancer, bitter as the cud
> Of vile, incurable sores on innocent tongues,—
> My friend, you would not tell with such high zest
> To children ardent for some desperate glory,
> The old Lie: *Dulce et Decorum est*
> *Pro patria mori.*

The arts, for one thing, permeate the Waldorf curriculum (and the first year in Waldorf teacher training is characterized by an immersion in the arts). Other "arts-based" curricula exist, of course, but they typically are not informed by the kind of clear anthropology and attention to the development of the human person from childhood to adulthood that undergird Steiner's project. Steiner's pedagogical anthropology, though his terminology often befuddles educators habituated to thinking in metrics, is amazingly sound. Like Maria Montessori, Steiner attends phenomenologically to the stages of dentition (change of teeth at age seven, arrival of the molars at twelve, appearance of the wisdom teeth at nineteen) as keys for understanding the development of the capacities for thinking and abstract thought in the child, though his picture of human development is more comprehensive and developed than Montessori's. And, unlike Montessori, Steiner recognizes the role of imagination in education.

Furthermore, Waldorf is an inherently Christian education, though not externally connected to any denomination. This Christianity does not start with a simple rehearsal of catechetical statements, nor with the "coloring-book" approach to early childhood religious instruction. It starts with the development of gratitude in the child:

> Love for all creation must be added to this gratitude. And if we lead the child on to the age of nine or ten in the way described, nothing is easier than to reveal in the living world around him the qualities he must learn to love. Love for every flower, for sunshine, for rain—this again will deepen perception of the world in a religious sense. If gratitude and love have been unfolded in the child before the age of ten, we can then proceed to develop a true sense and understanding of duty. Premature development of the sense of duty by dint of commands and injunctions will never lead to a religious inwardness.... We must show the child how divinity is immanent in all nature, in all human evolution, how it lives and moves, not only in the stones, but in the hearts of other men, in their every act.... In this way, the basis for a right attitude to the

> Mystery of Golgotha just between the ninth and tenth years is laid down.[10]

How superior such a gentle approach is to the soul-destroying power of *The Baltimore Catechism* and its arid progeny. It is my claim that premature intellectualization of the Faith (as done so well by most Christian religious formation programs—Protestant and Eastern Orthodox as well as Catholic—upon the model of *The Baltimore Catechism*) is more responsible for the emptying of the churches than are the lures of the world offered to young innocents exposed to college and popular culture.

We could use some more joy, more depth, as well as more life. All too often, the Catholic schools I visit are just as deadly (and deadening) as their public school counterparts. Likewise, too many Catholic or Christian homeschool groups I have visited are permeated by the "atmosphere of the bunker," as if the Gestapo will come by at any minute to put everyone on a train. This milieu is permeated by a spirit of anxiety. It is the furthest thing from the oil of gladness with which we were anointed. And it needs to go. We need a renewal of Christian life and Christian culture. The renaissance of apologetics is over (and has largely been a failure). Explanations are one thing; joy is another. And we are called to make a joyful noise.

Perhaps one of the most commendable aspects of Waldorf education is the joyful reverence with which Steiner schools observe the Christian year. I spent sixteen years as a Waldorf teacher, and it was in those communities that I first

10. Rudolf Steiner, *A Modern Art of Education*, 3rd ed., trans. Jesse Darrell (1928; repr., London: Rudolf Steiner Press, 1954), 179.

saw what it means to celebrate the Christian festivals, something all of my Catholic education failed to do. The list of festivals is elaborate, and Waldorf communities go to great lengths to make them truly celebratory for the children. The teachers, for one, perform a medieval play about the visit of the shepherds to the Christ child at Christmas. The schools have an Advent Garden (replete with "angels") for the kindergartners; the whole community celebrates Michaelmas (September 29), Martinmas (November 11), Santa Lucia's Day (December 13), May Day (May 1)—and Easter also. Some communities even come together in summer—when school is out!—to celebrate St. John's Day (June 24). And they are not even officially Christian (for the most part). My education was largely in Catholic schools—where I never heard of most of these festivals and where I certainly never felt the spirit of joy and celebration that characterizes the Waldorf calendar (which is what drew me to this education in the first place). And that joy is Christ himself. Imagine what would happen if Waldorf communities were actively involved with Him. They put every Catholic school I have ever seen to shame. They hold on to what Catholic education has lost: a Christian cosmological awareness.

"Cosmological awareness" is a fancy way to say that we need to attend to and interiorize the rhythms of nature and their relationship to the Church year. This would have different applications in the southern hemisphere than in the northern, of course, but, wherever one lives, it should not be all that difficult to connect the two. The Church year and the cycle of saints days and feast days enrich the soul in their reciprocity—though I am often mystified and disheartened by the reforms of the calendar in the Latin Church, not to mention horrified by "Ascension Sunday" and the other utilitarian modernizations by which holy days of obligation are transferred to the nearest convenient Sunday. I think just about everyone can feel how wrong this is, even at a bodily level. A sacramental sensibility arises from acknowledging the cosmological structure of our lives. By it we feel the presence of the Lord as his wisdom shines through the universe and our participation, for "from the mouth of the Most High [Wisdom] came forth, and covered the earth like a mist" (Sirach 24:3). How much richer, then, can be the experience for those of us privileged with access to the sacraments? We need to celebrate these interstices of the wheel of the year and the liturgical cycle—and in our communities. Otherwise, we are cut off from the source of life (*zoë*) and have no life in us (John 6:53).

Unfortunately, Waldorf education, like biodynamic agriculture (another of Steiner's contributions), though growing in popularity, has been cursed with boutique-ification as a school for the children of cultural creatives, tech gurus, and yuppies. It has been relegated to "exotic status," which is not at all what Steiner had in mind. Likewise, Waldorf has often capitulated to ephemeral cultural zeitgeists—whether to standardized testing, selling itself as "multicultural" (ah, the nineties!), or surreptitiously countenancing various progressive causes (for example, hormone therapies because gender is "culturally constructed")—thus selling out Steiner's commitments for a dish of popularity lentils. A good many Catholic and other ostensibly Christian schools, of course, have done much the same. It's like a virus.

But I'm not interested in introducing "Waldorf, the Catholic Version." As much as I admire Steiner and his educational gift, absorption and regurgitation is not a model I'd like to follow. Rather, what I have in mind is best described as a "postmodern sophiological hedge school," and could work for early childhood through college age, even beyond.

But it is best to not think of it in terms of a "school system" or "degree-granting institution," and even the word "school" could be misleading. It is much freer than that.

ANTHROPOLOGY

Finding a way to educate will prove fruitless, if not harmful, in the absence of a coherent pedagogical anthropology. "What is man that thou art mindful of him? and the son of man, that thou visitest him?" (Psalm 8:5). Unless one can answer this question, it is better to not try educating at all.

My own underlying assumption is that the Wisdom of God (Sophia) informs all of creation, the natural as well as the supernatural, the interior spaces of being as well as all chemical, biological, and cosmological processes. This all seems very obvious to me, and, until the rise of modernity, this assumption was universally held. However, with the rise of secularism, scientific materialism, and the pernicious influence of John Dewey's pragmatic educational paradigm of social reform (among many other things), this very essential pedagogical anthropology has disappeared from the cultural imagination, parochial schools *not* excepted.

Nevertheless, specters of this sophiological assumption persist, though primarily in a watered-down and mostly ineffectual lip service paid to "developmentally appropriate" methods of instruction. Maria Montessori, a physician by training, attended more phenomenologically (especially in her awareness of the importance of the stages of dentition) and so maintained at least a foothold in an anthropologically sound pedagogy. Likewise, Egan's stages of understanding, though not particularly tied to biology, do echo such a pedagogical scaffolding and are particularly incisive through his phenomenological and empirical insights based on *what children actually do* (as opposed to what theorists *think* they do), as far as understanding goes, in their journey through primary and secondary education as generally conceived.

Steiner, though, offers a more useful pedagogical anthropology for those interested in a sophiological (that is to say, Christian/sacramental) educational philosophy. Steiner's terminology, however, can be a little daunting to those unfamiliar with his thought. For some, it is downright weird. Steiner's anthropology speaks to the development of different "bodies," as children move through schooling: the physical body, the etheric body, the astral body, and the ego. This schematic might sound off-putting and esoteric, but it is in truth a modernization of a more classical/medieval/early-modern human anthropology.

According to this view, the physical body is that mode of being which we share with stones and minerals. The etheric body (called in the earlier literature "the vegetable soul") is characterized by that which we share with plants: being and growth. The astral body, which was in earlier times called the "animal soul," has being, growth, and desire or motivation as its attributes. Finally, the arrival of the constitution particular to the human person, the ego (Latin for "I") crowns all.[11] For Steiner, the etheric constitution, though always present in human life, predominates development from birth to about the age of seven (when the change of teeth occurs). Likewise, the astral body, though always present to some degree, arrives at puberty (the change in dentition with the arrival of the twelve-year molars indicat-

11. See, for example, Augustine's *Civitas Dei*, 7.29, as well as Aquinas's *Summa Theologiae*, 1.2.22 and 1.2.28.

ing a stage in this development). The ego, again, always present, comes into its own between the ages of nineteen and twenty-one (with the arrival of the wisdom teeth).

Central to Steiner's pedagogy is the attention paid to proper development of the physical, etheric, and astral bodies in order to facilitate a proper "birth" of the ego, the mode of expression of the individual person's spirit. Because of this anthropology, Steiner's pedagogy attends to the need for hygienic rhythm (in daily activities, in the week, the month, in repetition, in breathing through music and the recitation of poetry and rhymes, etc.) and an immersion in beauty between the ages of seven and fourteen, which, he argues, will strengthen the etheric health of the child before the astral forces arrive *en force*, potentially wreaking havoc on the bodily health of the child if the etheric is not strengthened and in a healthy state. He also prescribes work with real things—knitting, crochet, gardening, woodworking, music, movement (eurythmy), and so forth—in addition to more academic (though artistically explored) subjects. In this way, the entire human being is educated. Indeed, Steiner's anthropology is in many ways the most Catholic aspect of his pedagogy. I am unaware of any comparable anthropology in Catholic, Eastern Orthodox, or Protestant pedagogies, which usually follow Dewey in some regard if they are not simply relying on a purely catechetical approach.

For Steiner, the success of his pedagogical anthropology depends upon the "livingness" of the teacher. Above all, he sees education as an endeavor of human interaction (much that goes by the name of "online education," I am sure, horrifies him). If I can say anything about the current state of mainstream education (among which I include parochial and independent schools), it's that it is a deadening environment. Education should impart life. As Steiner spoke in 1923:

> We must bring life into the world, which through its dead intellectualism is faced with the danger of falling still further into death.
>
> Basically, we have fallen out of the habit of being inwardly incensed by things as they are. If you merely pull a long face, however, about things that ought not to be rejected in our civilization, you certainly will not be able to educate. . . .
>
> One might almost say that the correct symbol for institutions of higher education today would be a thick black pall hung somewhere on the wall of every lecture room. Then one would realize that behind it there is something that must not be shown, because to do so would throw a strange light on what goes on in these lecture rooms! Behind the black pall there should be a picture of Michael's battle with the dragon, the battle with deadening intellectualism.[12]

Things have not improved much during the intervening one hundred years.

Another aspect of a healthy anthropology—and not only a healthy pedagogical anthropology—can be found in revisiting and reimagining the very traditional (though discarded in even the most traditionalist curricula) understanding of the human person's intimate relationship with the cosmos, of the microcosm to the macrocosm.

The notion of the microcosm–macrocosm synergy can be found in nearly every anthropological and spiritual tradi-

12. Rudolf Steiner, *Deeper Insights into Education: The Waldorf Approach*, trans. René Querido (Spring Valley, NY: Anthroposophic Press, 1983), 49–50.

tion on earth. It was only with the arrival of modernity that it was kicked to the curb by those who "know better." Modernity's estrangement from nature can be directly attributed to this exile (which mirrors—and is really a part of—the simultaneously occurring exile of God), an exile postmodernity embraces. I would even go so far as to suggest that our current infatuation with cyberspace is an unconscious desire to recapture such a relationship. But cyberspace is itself an exile of the Real; such a desire, therefore, can never be fulfilled in that world, but can only further our estrangement from the Real. It's called "virtual reality" for a reason. It's a glittering cage of souls.

Simply put, the idea of the microcosm–macrocosm suggests that the "lesser world" of the human person mirrors and is implicit in the "greater world" of the cosmos and that the greater world is implicit in the lesser, an idea that David Bohm's concept of "implicate order" does much to verify.[13] In the history of the West, this relationship was the foundation for astrology. The positions of the planets before the constellations of the zodiac at one's birth, for example, were thought to correspond to and influence the individual's human biology: Ares = head; Taurus = neck; Gemini = limbs; Cancer = breast; Leo = heart; Virgo = digestive organs; Libra = pelvis; Scorpio = reproductive organs; Sagittarius = thighs; Capricorn = knees; Aquarius = shins; Pisces = feet. Likewise, in the process of incarnation the soul traversed the starry realms on its way to birth, a concept beautifully illustrated in Macrobius's "Dream of Scipio," related by Cicero and present in Dante as well. And as Pythagoras, Macrobius, and Robert Fludd (among others) have attested, this journey and the cosmos itself are characterized by a

13. See Bohm and Hiley, *The Undivided Universe.*

celestial music, as Shakespeare's Pericles marvels: "Rarest sounds! Do ye not hear?"

Anglican priest and poet Thomas Traherne (1636?–1674) echoes this sensibility in his poem "Wonder":

> How like an angel came I down!
> How bright are all things here!
> When first among his works I did appear
> O how their glory me did crown!
> The world resembled his eternity,
> In which my soul did walk;
> And ev'ry thing that I did see
> Did with me talk.[14]

Traherne's contemporary Nicholas Culpepper (1616–1654) explains the concept further:

> If you do but consider the whol Universe as one united Body, and Man an Epitomy of this Body, it will seem strange to none but Mad-men and Fools that the Stars should have influence upon the Body of Man, considering he being an Epitomy of the Creation, must needs have a Celestial World within himself; for to wind the strings a little higher, If there be a Trinity in the Deity (which is denied, but by none but Ranters) then must there be a Trinity also in all his Works, if there be a Unity in the God-head, there must needs be a Unity in all his Works.[15]

I am not suggesting, however, that we start practicing astrology or pay more attention to the horoscope page in the newspaper. The stars incline, brothers and sisters, they do not compel. What I am suggesting is that we start paying

14. *The Poetical Works of Thomas Traherne*, ed. Bertram Dobell, 2nd ed. (London, 1906), lines 1–8.

15. Nicholas Culpepper, *Pharmacopoeia Londinensis, or The London Dispensatory* (1654), A3v.

more attention (much more) to our own integration into the cosmos and the cosmos's integration into our own being. The alienation from nature that was born out of nominalism, became absurd with the debate over *natura pura*, gained traction with Reformed theology (to say nothing of Counter-Reform theology), and triumphed with the Scientific Revolution, dominates our current cultural imaginary. It is a poisonous, demonic, and unrepentant worldview that has no place in an education based on reality. An education without a living cosmology is not deserving of the name.

STRUCTURE

One thing I have learned about education, having homeschooled for twenty years and taught K–12, college, and university over the course of twenty-five, is that education as usually practiced is enormously inefficient with regard to time. This is most glaringly obvious in the K–12 model. What, for instance, is the rationale for an 8–3 (much later in the case of aftercare), Monday through Friday schedule? My best guess is that it functions much in the way of those lockers at a bus station: a container for storing one's valuables in a more or less safe place until they can be retrieved. This allows both parents (or the parent) to work, of course, and keeps the kinder off the streets, while more exclusive college-preparatory schools do their best to reify the master/slave dialectic that is Western civilization and is contaminated by both the capitalist and Marxist species of poison. But, really, this is not the healthiest of situations. Ivan Illich notes the inefficiency of the school model, arguing that children could learn anything offered in a given curriculum in a much shorter time (he seems to prefer drilling and immersion). But inefficiency is really by design, and Illich's diagnosis in this regard is as scathing as it is accurate concerning the "hidden curriculum" implicit in the educational franchise:

> Everywhere this . . . curriculum instills in the pupil the myth that increased production will provide a better life. And everywhere it develops the habit of self-defeating consumption of services and alienating production, the tolerance for institutional dependence, and the recognition of institutional rankings. The hidden curriculum of school does all this in spite of the contrary efforts undertaken by teachers and no matter what ideology prevails.[16]

He further exposes this project and its insidiousness in human society:

> Rich and poor alike depend on schools and hospitals which guide their lives, form their world view, and define for them what is legitimate and what is not. Both view doctoring oneself as irresponsible, learning on one's own as unreliable, and community organization, when not paid for by those in authority, as a form of aggression or subversion. For both groups the reliance on institutional treatment renders independent accomplishment suspect.[17]

Nevertheless, the system prevails. People love their cages.

Indeed, I have always admired how Laura Ingalls Wilder, Beatrix Potter, Frederick Douglass, and William Blake (and many, many others) were able to accomplish so much with their putatively limited or atypical schooling: Wilder in relative isolation on the plains and in the forests of frontier America, Potter with a tutor at home, Douglass on the sly under threat of severe punishment, and Blake at home in working-class London during the eighteenth century. As

16. Illich, *Deschooling Society*, 106.
17. Ibid., 3.

Blake would later write: "Thank God, I never was sent to school/To be Flog'd into following the Style of a Fool."[18] Furthermore, as Benjamin Constant wrote of his own education at home, a freer structure can make for a more congenial learning environment:

> The first tutor of whom I have a reasonably clear recollection was a German named Stroelin. . . . He had had the ingenious notion of getting me to invent Greek in order to teach it to me, that is to suggest that the two of us create our own language which only we could understand. I became enormously keen on the idea. First of all we devised an alphabet into which he introduced Greek letters. Then we began a dictionary in which each French word was translated by a Greek one. All of this imprinted itself on my mind to a remarkable degree because I believed I was the inventor of it all, and I already knew very many Greek words and was in the process of applying general rules to what I had made up—that is I was learning Greek grammar—when my tutor was thrown out of the house. I was 5 when that happened.[19]

Two questions remain, of course: 1) whether any school today would be hospitable to the method by which Constant learned Greek; 2) whether in our times Blake would be placed on medication on the first day of school, or the second. The answers are "no" and "first."

Another structural weakness in current models of education is twofold: an overemphasis on engorging students

18. "You say their Pictures well Painted be," *The Complete Poetry and Prose of William Blake*, ed. David V. Erdman (Berkeley: University of California Press, 1982), lines 3–4.

19. Quoted in Dennis Wood, *Benjamin Constant: A Biography* (London: Routledge, 1993), 14. Thanks to Kieran Egan for the reference.

with "content" (much of it empty and dripping with the bureaucratic paraphernalia of socialization) and the consequent absence of contemplation. Contemplation—by which I mean being present to phenomena and not necessarily working under religious obligation—should really be the centerpiece of any education. Unfortunately, the mania for standardized testing, specious and suspiciously designed "benchmarks," and governmental meddling in curricula have seriously compromised any possibility for a contemplative education. Meanwhile, public schools, under tremendous pressure to achieve sufficient test scores to ensure their survival, have no recourse but to dedicate their efforts to test preparation. This creates a stressful environment for both students and teachers—which further inhibits the creation of a congenial learning environment, let alone the possibility for a contemplative approach. Students under duress are unable to learn optimally, or, to use the telling jargon of the educational marketplace, they are unable to "perform." And let's admit it: this entire testing charade is nothing but a performance—political theater at its most despicable, with children as props.

Making room for contemplation, or what physicist and educator Arthur Zajonc calls "contemplative inquiry," means exactly that: making room, creating a space. An education crammed with content, activities, and rushing from one thing to the next allows no such space. This is a pathological situation. As Simone Weil describes it, contemplative attention "forms the real object and almost the sole interest of studies."[20] This kind of attention, argues Weil, even if to a

20. Simone Weil, "Reflections on the Right Use of School Studies with a View to the Love of God," in *Waiting for God*, trans. Emma Craufurd (New York: Harper Perennial, 1951), 57.

difficult algebra or geometry problem, eventually "will bear its fruit in prayer. There is no doubt whatever about that."[21] Gaston Bachelard extends this conception in his embrace of reverie as a way of knowing:

> One can also understand the great value in establishing a phenomenology of the imaginary where imagination is restored to its proper, all-important place as the principle of direct simulation of psychic becoming. Imagination attempts to have a future.... A world takes form in our reverie, and this world is ours. This dreamed world teaches us the possibilities for expanding our being within the universe.[22]

A contemplative engagement allows children to "enter into phenomena," whether it be poetry, a proof in geometry, work in optics, drawing, making music, woodworking, blacksmithing, or any subject. As Zajonc notes, this allows an intimacy with Things that often quite spontaneously brings about the feeling of reverence.[23] Contemplation, furthermore, is indispensable for disclosing the Wisdom of God (Sophia) that inwardly and outwardly permeates the cosmos. Without it, education is merely training and data-gathering.

A tremendous hindrance to forging any kind of change in education, however, lies in the existence of accrediting bodies and governmental regulations to which most existing educational institutions are enslaved (and this touches

21. Ibid., 58.

22. Gaston Bachelard, *The Poetics of Reverie: Childhood, Language, and the Cosmos*, trans. Daniel Russell (Boston: Beacon Press, 1969), 8.

23. See his "Cognitive-Affective Connections in Teaching and Learning: The Relationship between Love and Knowledge," *Journal of Cognitive-Affective Learning* 3, no. 1 (Fall 2006): 1–9.

K–12 as well as higher education). This enslavement, as in chattel slavery, is all about economics. Governmental funding is tied to schools acquiring the appropriate accreditation, which more and more looks like bad MBA philosophy applied to pedagogy. Private and parochial schools, though typically free of the burden of government funding and at least some of the regulations that accompany it, have nevertheless been infected by the accreditation bug—yet another indication of the insidious omnipresence of secular education's superego. Teachers and professors, feeling powerless to do anything to fight against this culture of byzantine bureaucratic absurdity, have more or less adopted a "go along to get along" ethos regarding this problem, though it compromises their integrity and agency by a thousand tiny papercuts (almost literally!). They shrug their shoulders, roll their eyes—and then obediently add another "core component" or "institutional learning outcome" to a syllabus that is already twenty-five pages long.

But the existence of this kind of funding (especially through subsidized loans in higher education in the United States) has artificially inflated the cost of education, and not just tuition. Because funding is assured, and because students need textbooks, the publishers of academic textbooks have exorbitantly inflated the prices of their products (and usually give the authors only a 1–3 percent royalty). Illich was onto this ruse in 1971:

> School programs hunger for progressive intake of instruction, but even if the hunger leads to steady absorption, it never yields the joy of knowing something to one's satisfaction. Each subject comes packaged with the instruction to go on consuming one "offering" after another, and last year's wrapping is always obsolete for this year's consumer. The textbook racket builds on this demand. Edu-

> cational reformers promise each new generation the latest and the best, and the public is schooled into demanding what they offer.[24]

Since then, the game has been refined to the point where students are turned into ATMs for the system (let's call it "Edu, Inc."), and they are then stuck with a degree that is nearly worthless in many cases owing to educational inflation and are simultaneously saddled with unbelievable debt. But, because of a demographic shift that finds fewer available students, schools compete for enrollment—replete with marketing campaigns and communications directors—such that administrators and enrollment coordinators enlist students less and less prepared for college. But, prepared or not, the loans are still assured—so why not! All this while administrators and university presidents receive increasingly generous salaries and instruction is more and more farmed out to part-time adjuncts, turning the teaching profession into a service industry more or less on par with working the deep fryer in the fast food business. Can a person of integrity participate in such a system with a clear conscience? I think not. But if the institutional educational system cannot be changed (and I don't see how it can), it can at least be subverted or short-circuited.[25]

24. Illich, *Deschooling Society*, 61.

25. Here I must commend Therese Schroeder-Sheker, founder of the Chalice of Repose Project, who has single-handedly rejected the prevailing paradigms in work with the dying and in education and forged a truly inspiring, truly human, approach to clinical work and hospice care.

THE POSTMODERN SOPHIOLOGICAL HEDGE SCHOOL

From the early modern period until well into the nineteenth century, rural Ireland was home to the educational phenomenon known as a "hedge school." In the hedge school, illegal prior to Catholic emancipation in 1829, itinerant teachers would travel the countryside offering clandestine instruction in reading, writing, arithmetic, Irish history and literature, as well as in Greek, Latin, and the native language. The curriculum was very plastic and could also include bookkeeping, navigation, physics, and other subjects.[26] Classes were not conducted near hedgerows, but usually in a barn or some other hospitable space. The community provided fees for the teacher, and the level of instruction was often very accomplished. As one state inspector reported in 1835:

> Amazed at the skill of the twelve year old boys in reading the new books, and considering the possibility that they were reciting from memory, I invited one of their number to read me a passage from the gospel of Saint Matthew. Evidently the child misunderstood me. He searched in his satchel until he found his tattered book, stood up, and proceeded to read me the account of Christ's passion—in Greek.[27]

Following emancipation, Catholic children eventually migrated to the parish schools or to the state schools, the latter being, for the most part, nonsectarian; the hedge schools

26. Yolanda Fernández-Suárez, "An Essential Picture in a Sketch-Book of Ireland: The Last Hedge Schools," *Estudios Irlandeses*, no. 1 (2006): 45–57, at 48.

27. Quoted in Fernández-Suárez, "An Essential Picture in a Sketch-Book of Ireland," 48.

dwindled and eventually disappeared. The hedge school model, however, offers us at least some possibilities as an example of sustainable noncooperation, practicality, and efficiency.

In the past, school as a place where "The children learn to cipher and to sing,/To study reading books and history,/To cut and sew, be neat in everything/In the best modern way"[28] could be enough. An emphasis on intellectual development made sense for school in a world where people—children as well as adults—were engaged in the Real in their everyday lives, such that intellectual development and training worked as a supplement to their being. In parish schools and in monastic communities, contact with the things of nature being already present in the lives of those involved—through raising animals, gardening, handwork (knitting, sewing, and so forth), woodworking, blacksmithing, and other crafts—intellectual training as the primary emphasis in schooling served as a balance. Likewise, whether in a formal sense as in monastic communities, or informally as in rural life animated by the intertwined realities of the Church and agricultural years, contemplation was already built into the structure of life. We all know that such is no longer the case, yet education as usually delivered persists in a highly intellectualized and competitive environment (Darwinian to the core) that rewards academic achievement (and, tangentially, in the spirit of capitalism, athletic ability) to the detriment of all other human capacities. This is so, no doubt, because the designers of modern education—professors, legislators, governmental agencies, accrediting bod-

28. William Butler Yeats, "Among School Children," in *The Collected Poems of William Butler Yeats* (New York: Macmillan, 1956), lines 3–6.

ies—have designed the system in their own images. That is to say that the system rewards those who are most like the people who designed it, or at least those whose investment in the system has become a sort of unconscious religion. It's a self-generating perpetual motion machine. The hedge school model, if sensibly implemented, could offer a healthy alternative to such bureaucratic madness.

The hedge school would create an environment in which the morning could be devoted to studies in the traditional sense (the three R's, languages), but, as Steiner demanded, they should be taught artistically. No textbooks. No computers (at least before the age of fourteen). No ugly and utilitarian classrooms filled with ugly and utilitarian furniture and an ugly and utilitarian curriculum. Only engagement with what is real: color and sound, beauty and presence, human interaction and contemplation. That means that students would need *time to think, time for reverie* in addition to time for instruction. Afternoons could then be devoted to developing practical and artistic skills. In the classical Irish hedge school, such would be redundant; in our own time, they are absolutely vital.

The recalibration of the human ego to participation in the Real is fundamental to my conception of the postmodern sophiological hedge school. There could be many varieties of hedge school, but without participation in the Real, they would only be adaptations of the educational superego that permeates the culture. My claim is that a way of learning imbued with the arts and engaged with practical activities combined with a contemplative ethos provides a corrective for the human soul seriously damaged by a culture characterized by the technology, isolation, and synthetic media that insulate human persons from nature, the cosmos, and, ultimately, from God. Steiner, of course, saw all this long ago, and the system he developed goes a long way toward bringing a corrective to a social imaginary gone wrong. However, though Steiner was deeply aware of the importance of Christ for humanity, Waldorf educators tend to be a bit more aloof from even his idiosyncratic emphasis on Christ and often hold those with more traditional understandings of Christianity in unbridled contempt. And aloofness from Christ is the paradigmatic distancing from the Real. A sacramental awareness, so implicit in Waldorf pedagogy, could be so much more effective through a more explicit commitment to sacramental facticity. Furthermore, Waldorf education, for all the good it contributes, is still overshadowed by (and therefore participates in) the educational superego. As Steiner explains, "The principles of Waldorf School education . . . are . . . in no sense revolutionary . . . . There is no desire to cast everything aside and imagine that the only possible thing is something radically new."[29] Steiner, for all his visionary talents, was perhaps unable to see how the prevailing educational paradigm influenced even his capacious imagination.

One of the most important features of a hedge school is that the parents would be the ones in charge. There would be no reason to attend school-board meetings to see if those august protectors of the state's investment in social engineering might condescend to parental concerns. Furthermore, the hedge school model would be absolutely flexible—no one would be held to an arbitrary "minimum number of school days" per year or to the K–12 conveyor belt. Students would learn at their own pace and more effi-

29. Steiner, *A Modern Art of Education*, 39.

ciently, not condemned to serving time, not moving around like Pavlovian hounds every time a bell rings. Parent collectives could find a teacher or teachers from among themselves or from outside, but they would all by design be committed to the project. They might, however, need to decondition themselves from a toxic notion—the result of their own indoctrination within the system—of what school or an education is. As I have seen in far too many homeschool co-ops, for instance, the idea that school is an us-versus-them (teachers-versus-students) environment readily appears, as do all of the other school stereotypes found in the cultural imaginary. The egregore of institutional education is a pernicious and strong one. And we have been so well trained.

As is all too obvious, schooling as usually practiced encourages all manner of pathologies—bullying, envy, shaming, violence, not to mention, alas, murder. It also encourages a false perception of values—whether or not the state's socialization needs are behind this encouragement. Indeed, this false perception is more structural than intentional. As Illich argued,

> If there were no age-specific and obligatory learning institution, "childhood" would go out of production. The youth of rich nations would be liberated from its destructiveness, and poor nations would cease attempting to rival the childishness of the rich. If society were to outgrow its age of childhood, it would have to become livable for the young. The present disjunction between an adult society which pretends to be humane and a school environment which mocks reality could no longer be maintained.

To which he adds: "The disestablishment of schools could also end the present discrimination against infants, adults, and the old in favor of children throughout their adolescence and youth."[30] Childhood, Inc. even touches the more or less wholesome Waldorf movement—with "Waldorf dolls" selling for over $100 and other "Waldorf toys" similarly priced (so counter to Steiner's intentions).

REIMAGINING EDUCATION

It should be clear that changing our perceptions concerning the being of education is inextricably linked to changing our perceptions about how we live in general, that is, in terms of *oikonomia*. I don't pretend to think that what I propose as a postmodern sophiological hedge school will have much if any societal impact, not that I'm concerned that it should. The classical Irish hedge schools worked because the communities who adopted them saw a need; and they disappeared when that need disappeared. I also see a need for a paradigm shift at our cultural moment, but postmodern life is so exceedingly crammed with diversion, busyness, and distraction, that many people, even if they see a need, may feel powerless to do anything about it. We are paralyzed by a postmodernity that more and more appears an ultramodernity. I understand that. Such paralysis, however, comes at a cost: the tacit and mostly unconscious acceptance that makes people "like their inescapable social destiny,"[31] robbing them of all save the illusion of agency. But there comes a time when people finally say, "Enough." Saying "enough" when it comes to education, however, quickly leads to saying the same thing when we consider

30. Illich, *Deschooling Society*, 41.
31. Aldous Huxley, *Brave New World*, 16.

how our lives are ordered in every sphere and by whom or by what. That may be too much for some people. Good.

4

OIKONOMIA: THE HOUSEHOLD OF THINGS

(POST)MODERNITY'S primary sin resides in not knowing what life is. This is as true in economics and politics as it is in the arts and sciences. Our attention adheres to *bios*, and we lack the proper organs of perception for discerning *zoë*. Unless this can be remedied, there is no hope for such a thing as a Catholic culture.

(Post)modernity knows what "making a living" (*bios*) is. We ascribe value (consciously or unconsciously) to what a person *does*—lawyer, farmer, mechanic, trash collector. Indeed, Marxist and even Maoist critique (and the critiques of feminism, postcolonialism, etc.) has focused primarily on precisely such locators, as if value has something to do with power, profit, or productivity, which only implicates them all the more in the capitalism they so righteously seek to discredit (as has been made only too obvious by the ascendance of China as an economic power). The Internet and social media, in which everyone by participation or association becomes a brand or product, have only exacerbated this ethos. Thus *bios*.

Zoë, on the other hand, lies beyond us, even for those self-identifying as Christians. For with Christians, too, the focus is all too often on what one *does*—e.g., the praxis of the Traditionalist, participation in a megachurch, advocacy of social justice—to the detriment of the life that supposedly inheres faith and that was promised by Christ. *Bios*, of course, is important—and inescapable—but only to the degree to which it can reveal *zoë*. If *bios* cannot reveal *zoë*, we really have no life in us at all (John 6:53).

Participation in the sacraments, clearly, is by definition participation in the *zoë* of Christ. However, we must admit that many who do participate in the sacraments do not simultaneously disclose the *zoë* they impart. This, I would argue, has to do with our cosmological disengagement. Our sacramental lives, for the most part, are confined to Sundays, and primarily between 10:00 and 11:00 a.m. Outside of that precious hour, we are most of us as ensconced and implicated in a materialistic, disenchanted world as any hardened philosopher. Descartes himself was a pious believer. So are we all.

The trajectory by which we were delivered from engagement with the cosmos extended over a long period of time and proceeded incrementally. Even though the theologians of the Middle Ages sought to continually refine doctrine and systematics, their progress generally did not impact the faithful, for the most part illiterate, who maintained folk customs alongside or integrated with the liturgical cycle. Though upon occasion this or that reformer sought to curtail what were thought of as superstitious practices from the lives of the common people, in general high theology and folk sensibility lived in relative harmony. During the Renaissance, however, theologians exerted more effort in creating what Charles Taylor has described as a "one-speed religion" that sought to "purify" Christendom of "superstition."[1] Erasmus, for instance, dismissed pilgrimage, the veneration

1. Charles Taylor, *A Secular Age* (Cambridge: Harvard University Press, 2007), 62–63.

of relics, praying to saints, and even images as evidence of the "ocean of superstition" to which the faithful were inclined, upon all of which Protestant reformers capitalized.[2] One can easily surmise what he thought about maypoles, jumping over St. John's fires, and belief in faerie. Ostensibly a quest for uniformity, in truth Erasmus's reform project was an effort by elites to whip the peasantry into shape: their own.

Prior to the early modern period, the Christian year, at least in the Western Church, was characterized by an abundance of feast days, which were in turn characterized by both piety and carnival in mixed and multifarious measure, a habit which caused bishops Walter de Cantelupe and Robert Grosseteste (among many others) some consternation in the mid-thirteenth century.[3] However, with the Protestant Reform, many of these feast days, even Christmas for a time, disappeared, and along with them their respective and particular aspects of pious and riotous conviviality. In the early years of the Henrician reforms in England, for example (early 1530s), the House of Commons reduced the number of feast days to a bare minimum, maintaining only feasts of the Apostles and the Virgin Mary, though with the Edwardian and Elizabethan reforms even these disappeared from the calendar and civic life.[4] The Council of Trent continued in this direction in its own "Catholic" way[5] (a prac-

2. Carlos M.N. Eire, *War against the Idols: Worship from Erasmus to Calvin* (Cambridge: Cambridge University Press, 1986), 36–45, at 37.

3. Ronald Hutton, *The Rise and Fall of Merry England: The Ritual Year, 1400–1700* (Oxford: Oxford University Press, 1994), 56.

4. See Hutton, *The Rise and Fall of Merry England.*

5. John W. O'Malley, introduction to *Collected Works of Erasmus: Spiritualia, Enchiridion, De Contemptu Mundi, De Vidua Christiana*, ed. John W. O'Malley (Toronto: University of Toronto Press, 1988), xxxii.

tice that was furthered by the "reformation" of the liturgical calendar by Vatican II and the manner by which it expunged a considerable number of saints from the *Communio*). Some folk customs nevertheless persisted for a time as palimpsests of an earlier age, evidence for which is found, for example, in the poetry of Robert Herrick (1591–1674), wherein the Anglican priest elegizes rural England's Catholic folk religion heritage.[6] But over time most of these traditions evaporated into the ether.

The question I would like to pose, then, is whether a sacramental life is even possible divorced from a cosmological awareness. At best, such would resemble a version of *natura pura*. And that's not saying much, since *natura pura* is the poisoned cup to which (post)modernity has developed an addiction and which "opens out the space of secular autonomy itself, within which all transcendent reference will eventually be denied."[7] My claim is that a sacramental life separated from a cosmological awareness cannot be construed a proper sacramental life. And life is not life, if it's not sacramental.

An accompanying—and by no means unrelated—secular development was the gradual removal of the very fundamental human relationship to the land, which further alienated culture from cosmology. This occurred through the drawn-out (over a few centuries) progression of enclosure

6. See the chapter entitled "Robert Herrick, Little Gidding, and the Spectres of an Old Religion" in my *The Incarnation of the Poetic Word: Theological Essays on Poetry and Philosophy / Philosophical Essays on Poetry and Theology* (Kettering, OH: Angelico Press, 2017), 67–93.

7. John Milbank, *Beyond Secular Order: The Representation of Being and the Representation of the People* (Malden, MA: Wiley Blackwell, 2013), 4.

laws in Europe (and eventually throughout the West) that removed the peasantry and the poor from scratching out a living from common lands and simultaneously removed them from their engagement with the cosmos. Whatever one's heritage, we are all their psychological and cultural descendants.

Thomas More (1478–1535) laments the early implementation of enclosure in *Utopia* (1516) and the rapaciousness of those avaricious nobles who took away the meager livelihoods of the peasantry:

> Now they are becoming so greedy and wild that they devour men themselves, I hear. They devastate and pillage fields, houses, and towns. For in whatever parts of the land the sheep yield the softest and most expensive wool, there the nobility and gentry, yes, and even some abbots though otherwise holy men, are not content with the old rents that the land yielded to their predecessors. Living in idleness and luxury, without doing any good to society, no longer satisfies them; they have to do positive evil. For they leave no land free for the plow: they enclose every acre for pasture.[8]

Such greed not only left the peasantry without a living, but often left them in their desperation as further victims of a cruel and rigged law:

> They leave the only homes familiar to them, and they can find no place to go. Since they cannot afford to wait for a buyer, they sell for a pittance all their household goods, which would not bring much in any case. When that little money is gone . . . what remains for them but to steal, and

8. Sir Thomas More, *Utopia*, trans. and ed. Robert M. Adams (New York: W. W. Norton & Company, 1992), 12.

> so be hanged—justly, you'd say!—or to wander and beg? And yet if they go tramping, they are jailed as sturdy beggars.[9]

In the mid-seventeenth century, Gerrard Winstanley railed against the perpetrators of such an unjust system in prophetic terms:

> The earth was not made purposely for you, to be the Lords of it, and we to be your Slaves, Servants, and Beggars; but it was made to be a common Livelihood to all, without respect of persons: And this buying and selling of Land, and the Fruits of it, one to another, is *The cursed thing*, and was brought in by War; which hath, and still does establish murder, and theft, in the hands of some branches of Mankinde over others which is the greatest outward burden, and unrighteous power, that the Creation groans under: For the power of inclosing Land, and owning Property, was brought into the Creation by your Ancestors by the Sword; which first did murther their fellow Creatures, Men, and after plunder or steal away their Land, and left this Land successively to you, their Children.[10]

Three hundred years after More and nearly two hundred after Winstanley—following an encroaching and almost totalizing (and completely "legal") project of enclosure and cruelty—the poet John Clare (1795–1864) would lament, "Inclosure like a Bonaparte let not a thing remain,/It lev-

9. Ibid.

10. [Gerrard Winstanley], *A Declaration from the Oppressed of England, directed to all that call themselves, or are called, Lords of Manors, through this Nation; That have begun to cut, or through fear and covetousness, do intend to cut down the Woods and Trees that grow upon the Commons and Waste Land* (1649), A2r. Emphasis in original.

elled every bush and tree and levelled every hill/And hung the moles for traitors."[11] Clare, a farm laborer, suffered first-hand from enclosure and the poverty and subsequent malnutrition it caused: all of which doubtless contributed to the madness into which he gradually descended.

Some economists have argued that enclosure was sound economic policy.[12] E.P. Thompson dismisses such prattle and condemns enclosure for what it was: "a plain enough case of highway robbery, played according to fair rules of property and law laid down by a Parliament of property-owners and lawyers."[13] So let someone take away the land of the defenders of enclosure in order to turn it to greater profit and then see what they think. Enclosure, however, was not only a tool of capitalist avarice: Soviet collectivization also employed its own version of the practice.[14] And it never really went away; it only goes by different names, such as "eminent domain." But this avaricious practice is not typically manifested through such violent means. John Taylor Gatto, for example, has argued that current real estate practices in the United States which divide and conquer rural lands to optimize capital through gentrification—often raising taxes 600 percent or more over time—also squeeze

11. "By Langley Bush," lines 7–9. Quoted in Jonathan Bate, *John Clare: A Biography* (New York: Farrar, Straus and Giroux, 2003), 50.

12. See, for example, George W. Grantham, "The Persistence of Open-Field Farming in Nineteenth-Century France," *Journal of Economic History* 11, no. 3 (September 1980): 515–31.

13. E.P. Thompson, *The Making of the English Working Class* (1963; repr., New York: Vintage Books, 1966), 218.

14. See Joseph Stromberg, "English Enclosures and Soviet Collectivization: Two Instances of an Anti-Peasant Mode of Development," *Agorist Quarterly* 1, no. 1 (Fall 1995): 31–45.

people off of their land and send them off to the "work-houses" of factory and retail "employment."[15] All by design.

A relationship to the land, and thereby to the seasons, naturally fosters a relationship to the cosmos, but such, in our technology-driven age, should not be assumed. Technology increasingly isolates us from the cosmos by feeding to us, especially through the Internet and smartphones, a *false sense of cosmos*. Indeed, the practices of industrial agriculture have done much to estrange the soil itself from the cosmos. Estranged from the real cosmos, we are ensnared in a virtual one. And we no longer remember how to discern the Real.

Indeed, the Internet is our commons. But unlike the rural commons of the past, the virtual commons disconnects us from the Real. In fact, like the elites who stole the commons from the peasantry, today's elites prefer to keep us in the virtual commons so we don't notice how we've been buff-ered from the Real and thus continue to consume like the sheep in More's *Utopia*, thence to be fleeced by our manag-ers. We turn to the virtual commons to fill a hunger that can only be satisfied by the sacramental cosmos. The virtual commons, it is true, fills, but it can never nourish. As a result, we waste away, malnourished, on a diet designed by demons.

To counter this a-cosmological (or, to be more honest, anti-cosmological) sterility, it will be necessary to re-pagan-ize Christianity. G.K. Chesterton, in his great wisdom, knew this to be true. "If we ever get the English back onto English land," he wrote in 1908, "they will become again a religious

15. John Taylor Gatto, "Guerilla Curriculum," *The Link: Homeschool News Network* 6, no. 2, homeschoolnewslink.com (2006). Accessed 18 April 2018.

people, if all goes well, a superstitious people. The absence from life of both the higher and lower forms of faith is largely due to a divorce from nature and the trees and clouds."[16] For, "take away the supernatural, and what remains is the unnatural."[17] Indeed, Catholic Christianity fulfills "the best pagan impulses."[18] Unfortunately, in recent years some American dioceses of the Roman Catholic Church have adopted the barbaric practice of transferring major feast days—Ascension and the Assumption of Mary, for example—to the following Sunday. The USCCB doesn't even wince. It truly doesn't get any more a-cosmological than this variety of practical nominalism.

People desire, innately desire, connection with the cosmos, which is at heart the sacramental desire for union with God, to finally know where we belong. Wicca and such attempts at nature worship can really be understood in precisely this way. But what they really desire, I would argue, is a Christianity not compromised by Enlightenment sensibilities, nominalism, and politics at the expense of mystery, reality, and beauty. Rosemary Haughton located these qualities in Celtic Christianity and "its feeling for the presence of the divine in natural beauty and in the unexpected, its single-minded devotion to the quest for God, and its sheer wealth of symbol and fantasy."[19] She recognized this same longing in the young people visiting the annual Glastonbury Festival. These young pilgrims, she observed,

16. G.K. Chesterton, *Heretics* (New York: John Lane Company, 1908), 101.

17. Ibid., 99.

18. John Milbank, *Theology and Social Theory: Beyond Secular Reason*, 2nd ed. (Malden, MA: Blackwell Publishing, 2006), xiv.

19. Rosemary Haughton, *The Catholic Thing* (Springfield, IL: Templegate Publishers, 1979), 105. My emphasis.

> earnestly studying the I-Ching, smoking hash, reading Thomas Merton, making candles, and discussing organic farming or the Second Coming are on a quest as much as young Parsifal. Like him they are apt to miss their chances, ask the wrong question, or not ask, and make the confusion worse. But at least they are setting out from their round table, and a few may achieve the Grail. *It is a thoroughly Catholic quest*, under the occasionally erratic guidance of Sophia. If, most of the time, they do not pay much attention to Mother Church, Sophia will no doubt help them to make contact with her when necessary.[20]

Many of them do eventually wander into the Church, but they are just as likely to wander out again. The community they hoped to find often proves a mirage, and they become dissatisfied or even feel betrayed in some way. The reality doesn't live up to the brochure. All they are offered is the intellectual purgatory of RCIA and a box of envelopes. Some stick around for a little while before moving on to fill their still unquenched thirst elsewhere. And, even if they stay, their children will probably leave. We don't have a culture that can hold them.

Christianity is at its foundations a pagan religion, especially if we take into account the etymology of the Latin *paganus* ("villager, rustic"). Simone Weil, more melancholically perhaps, thought Christianity a religion of slaves (though in a rhetorical context quite alien from Friedrich Nietzsche's). She was not incorrect, no doubt, but "pagan" is more all-encompassing a descriptor. The Apostles and disciples were almost to a person rustics. Over time, of course, Christianity became formalized and developed an elite of its own. I suppose this is just what happens. Nevertheless, I

20. Ibid., 130.

cannot help but recall the story of St. Francis of Assisi returning to his community after traveling to Egypt in his nearly successful attempt to convert the sultan. In his absence, they had "reorganized" their community, bringing in new furniture and a new administrative structure, generally modernizing and streamlining things. As his brother friars dined in newfound sophistication, Francis, dismayed, sat on the floor, covered himself and his food with ashes, and refused to participate. The administrative class, I am told, has a purpose. Only, I'm not sure what it is. But I do know that if Christianity were primarily a religion for a managerial class of intellectuals and aesthetes, it wouldn't be worth a damn.

The foregoing is no doubt why John Milbank has observed, and accurately, that now "the family, local economies, even the economy as a whole, along with voluntary associations, are treated as if they were politically incidental."[21] This situation may be expedient for the managerial class, but cannot be sustainable over the long run. The managerial class, seriously buffered from *zoë*, maintains a *bios* not only virtual, but at best as artificial as a plastic flower, at worst positively demonic. The unreal has replaced the Real. This is the source of the general malaise that permeates culture.

The ways by which we are buffered from the Real are manifold: from the environments in which we live and work, to the foods we eat, to the clothes we wear, to our understandings of health, our conceptions of nature, and our passive acceptance of what constitutes currency and exchange. Taking inventory of the synthetic accoutrements of our lives may prove disconcerting. Scientific breakthroughs continu-

21. Milbank, *Beyond Secular Order*, 156.

ally introduce new substances into our lives—fiberglass, plastics, polar fleece, GMO foods, pesticides, fungicides, innumerable varieties of drugs—but the scientific and political spheres (encouraged by business) rarely have the foresight to see what damage these "breakthroughs" might (and often do) cause, and new breakthroughs to repair the damages caused by the former breakthroughs are called for. This is madness.

Overwhelmed, we can do nothing about our condition other than make the best of a bad situation, or so it seems. But perhaps there *is* something we can do.

AS IF

In his fascinating book *The Time That Remains*, Giorgio Agamben observes that St. Paul's mission was grounded in founding "the *ekklēsia*, the messianic community."[22] Though many Catholic parishes and congregations of other denominations have adopted the milquetoasty rhetoric of calling themselves "communities" at the expense of "parishes," one has to wonder why the messianic vocation has been ignored in favor of "mission-statement Christianity." For either we are parts of a messianic community or we are just another social identity group—which is to be nothing. "The Messiah has already arrived, the messianic event has already happened, but its presence contains within itself another time, which stretches its *parousia*, not in order to defer it, but, on the contrary, to make it graspable."[23]

22. Giorgio Agamben, *The Time That Remains: A Commentary on the Letter to the Romans*, trans. Patricia Dailey (Stanford: Stanford University Press, 2005), 22.

23. Ibid., 71.

Agamben meditates on Paul's advice to his flock to live "as if" both in the world and not in the world; as if in time, yet also outside of time; as if a slave, yet also free. Is not this our only vocation? Yet how often we find ourselves only in our world, time, and condition, unable to recognize the presence of the Kingdom opened to us. In operational time, which it seems is the only time that really matters to us, we exist in a scientistic, materialist, utilitarian, and moralistic milieu defined by functionality and efficiency, the only respite from which we find in the Eucharist and in prayer, if we are inclined to participate in them.

To participate in the Kingdom is to participate in the Real. We must, therefore, participate *as if* the Kingdom existed: *because it does*. It truly awakens only when we are attentive to it. When we divert our gaze it disappears. To be attentive to the Kingdom is to be immersed in the oikonomia of things. My claim is that to participate in the Real is simultaneously to participate in the Kingdom, which is to already participate in the Church. This may or may not include formal membership in an ecclesial body, but we have to trust that Christ is the Real, the Kingdom, and the Bridegroom, or why even bother?

This true oikonomia will have many different forms and permutations. "Economy," as typically practiced in our culture at this historical moment "*is a function of death*.... However far man goes in his economic progress, he remains a slave, subject to death, even as he becomes a master."[24] The true purpose of oikonomia, according to Sergei Bulgakov, "is to defend and to spread the seeds of life, to resurrect nature. This is the action of Sophia on the universe in an effort to restore it to being in truth.... Economic activity overcomes

24. Bulgakov, *Philosophy of Economy*, 73–74. Bulgakov's emphasis.

the divisions in nature, and its ultimate goal . . . is to return the world to life in Sophia."[25] Our economic activity, however it is expressed, should be directed to this end.

A sound oikonomia, first of all, begins with a sound ontology, "a clarification of metaphysics."[26] Without a proper understanding of reality, we already find ourselves indentured to the unreal. Economist Guido Preparata suggests that such a wholesome oikonomia constitutes a "reinvention of consumption," and consumption, quite poignantly, begins in our relationship to nature. Our current relationship to nature through consumption, however, is characterized by pathology: Big Agriculture, for one, has done far more to envenom the environment than even General Motors; and the poisons with which Big Ag seasons our foods are practically impossible to avoid.

A proper relationship to nature, on the other hand, would compel us to *imitate* nature. As Preparata argues, "We must imitate nature by reproducing its essential processes—processes which are founded on three principles: zero-waste, solar energy, biodiversity & symbiosis."[27] Oikonomia, therefore, begins with agriculture, but not the toxic leviathan that is the agriculture we have come to tolerate if not accept. "The farmers and agriculturalists of our project," explains Preparata,

25. Ibid., 153.

26. E.F. Schumacher, *Small Is Beautiful: Economics as If People Mattered* (New York: Harper & Row, 1973), 86.

27. Guido G. Preparata, with Domenico D'Amico, Flavio Fabiani, Aurelio Riccioli, and Sebastiano Scrófina, "'The Blueprint': A Modest Monetary and Organizational Proposal for Re-launching the Economic Welfare of Communities," in *New Directions for Catholic Social and Political Research: Humanity vs. Hyper-Modernity*, ed. G.G. Preparata (New York: Palgrave Macmillan, 2016), 5.

> inhabit a landscape governed by these principles: the ideal model is that of the *bio-dynamic farm*. A bio-dynamic farm is characterized by the absence of importation (viz. closed-circuit cycle), zero-waste (the output of a sector serves as the input of another), diversity (crop-rotation and diverse ecosystems instead of intensive mono-cultivation), and a symbiotic relationship with all the elements of the wider living system.[28]

Such an approach is entirely congruent with what E.F. Schumacher sets out in his revolutionary text *Small Is Beautiful*. As Schumacher explains, "If we could return to a generous recognition of meta-economic values, our landscapes would become healthy and beautiful again and our people would regain the dignity of man, who knows himself as higher than the animal but never forgets that *noblesse oblige*."[29]

Over the past thirty years or so, the CSA (community-supported agriculture) movement has shown that not only is such an idea possible, but also financially feasible. Likewise, "urban farms" in Detroit and other cities have shown that land can be redeemed through organic and sustainable methods and bring fecundity to otherwise deadened landscapes and communities. St. Hildegard of Bingen called this quality *viriditas*, "greening power," and it characterizes all of nature, though often obscured or latent. And, lest we forget, human beings, too, are part of nature. The CSA movement brings people into touch with their own farmers, and they build a relationship not only with the farmers but with the land and the creatures which upon it dwell. So doing brings healing to the entire economic system, moving it ever closer

28. Ibid. Preparata's emphasis.
29. Schumacher, *Small Is Beautiful*, 109.

to Bulgakov's notion of the sophianic economy that "establishes the ultimate connection of all things."[30]

Unfortunately, the current economic systems available to us—capitalism, various strains of communism, socialism—are not very interested in the "ultimate connection of all things." Instead, they prove to be theme and variation on this or that iteration of economic Darwinism, characterized by binary thinking (some version of "us vs. them") and a restrictive ethos. These are not holistic paradigms, and the grand promises of their theorists and champions are never fulfilled, and never can be. When I was a schoolboy in the 1970s, for example, the prevailing thought was that automation and the aid of computer technology would within my lifetime create more leisure time for the working class, and full-time employment would consist of about three days of labor a week. This would create full employment and a prosperous society. Instead, we have automation and computer technology and people can't make ends meet at even sixty hours a week, while investors and shareholders reap the rewards that were promised to the working class. And the rewards are there. As Pope Francis has put it, while Marxism is wrong, the capitalist promise of prosperity through trickle-down economics is a ruse: "The promise was that when the glass was full, it would overflow, benefitting the poor. But what happens instead, is that when the glass is full, it magically gets bigger and nothing ever comes out for the poor."[31]

30. Bulgakov, *Philosophy of Economy*, 155.

31. Andrea Tornielli, "Never Be Afraid of Tenderness" [interview with Pope Francis], *La Stampa*, http://www.lastampa.it/2013/12/14/esteri/vatican-insider/en/never-be-afraid-of-tenderness-5BqUfVs9r7W1CJIMuHqNeI/pagina.html, 13 December 2013, accessed 29 April 2018.

Elsewhere, enthusiasts for distributism, a noble enterprise in theory, have raised their voices upon occasion in a meager but growing choir, but distributism itself has suffered substantial collateral damage in the dialectic between communism and capitalism. In general, distributism has been caricatured as an adolescent project appealing primarily to those attracted to the life of a hobbit. This is certainly unfair, but far too often writers and theorists of distributism lean too heavily on Tolkien (as well as Chesterton and Belloc) and often lapse into preciously styling themselves "monarchists" in a mimetically medieval cosplay. Though useful as self-fashioning or as an Internet avatar, this kind of distributism gives plenty of ammunition to those who would dismiss distributism as a romantic fantasy for not-quite grownups.

Aside from this image problem, the challenge for any attempt at distributism, however noble and good, is that it (at this point in history) can transpire only within the contexts of monetary systems already corrupted: and this is as true for the communist and socialist contexts as it is for the capitalist. And none of these systems will ever give distributism room to breathe and grow.[32] Not ever.

Nevertheless, all is not without hope. The recent explosion in cryptocurrencies (such as Bitcoin) indicates that the world monetary system can no longer claim absolute financial hegemony. And though Bitcoin can be used for nefarious purposes, other options are possible. As Preparata, drawing on the economic theories of Silvio Gesell (1864–1930) and Rudolf Steiner (1861–1925), has been arguing for

32. See, for example, Charles Heying, "Autonomy vs. Solidarity: Liberal, Totalitarian and Communitarian Traditions," *Administrative Theory & Praxis* 21, no. 1 (March 1999): 39–50, especially 48–49.

several years, the time for a *perishable currency* may at last be at hand. "If perishable money," he argues,

> which carries the anti-hoarding device in the expiration date, were injected into the productive fabric of society, it would outflank the banking network by spurring a circuit of its own—one where banks would on the one hand inevitably, and justly, surrender a sizable measure of decisional clout to the productive sector, and on the other, no longer base their investment policies on mere interest-driven exigencies.[33]

This is not to say that a perishable currency would usher in a utopia of communitarian and distributist fecundity, but it couldn't hurt. Money, like water, becomes rank and fetid when it sits in one place for too long; increasing its flow would help it to retain its vitality and allow it to impart health to the greater community. But, on the other hand, the archons of the world-banking and corporate-military-industrial complex would never let such a development, should it happen, flourish without a fight. We already know this. And the armies of Sauron would be nothing compared to the legions at the command of the archons, for they are many.

For these and other reasons, Preparata and others wonder whether such a hope cannot be realized until the system currently in power finally atrophies under the weight of its own greed and collapses. My concern, however, is not so much with tearing down, but with building up. It seems to me that people can start living a sophianic oikonomia *now*, however

33. Guido Giacomo Preparata, "Of Money, Heresy, and Surrender, Part I: The Ways of Our System, an Outline, from Bretton Woods to the Financial Slump of 2008," *Anarchist Studies* 17, no. 1 (2009): 18–47, at 20.

imperfectly realized. Our concern, that is, should be not the problems that now surround us but rather what we can do despite the restrictions brought on by the circumstances within which we find ourselves. Our focus is only one: to participate in the sophianic connection of all things and to nurture an oikonomia reflective of it; for this is the household of all Things.

Nevertheless, the daunting prospect of such a magnitude of change might cause even the stoutest heart to quail. This is especially the case considering the "massive resistance respectively opposed by our societies' academes, public habits and vested interests."[34] (Can we imagine an NPR segment on perishable currency?) Such a project does not need to be realized all at once, but could be implemented gradually as people more and more respond to an *as if* approach to our circumstances. If we were to live *as if* a sophianic oikonomia were a reality, even while we live in a world and are surrounded by a culture oblivious if not hostile to such an idea, the sophianic oikonomia would nevertheless come into being: it would have being, as it already has being. Only, by our attention to it, we would awaken it from slumber. Indeed, the call to economic activity, when considered in this light, is not the call to domination and exploitation, but the call "to return the world to life in Sophia."[35] And all of us privileged to have been baptized in the Name of the Holy Trinity are thus called; there is simply no other way. The responsibility terrifies, as an angel terrifies, but we are summoned nonetheless:

34. Guido Giacomo Preparata, "Of Money, Heresy, and Surrender, Part II: A Plea for a Regional and Perishable Currency," *Anarchist Studies* 18, no. 1 (2010): 8–39, at 35.

35. Bulgakov, *Philosophy of Economy*, 153.

I have heard that voice many a time when asleep
and, what is strange, I understood more or less
an order or an appeal in an unearthly tongue:

day draws near
another one
do what you can.[36]

36. Czeslaw Milosz, "On Angels," in *New and Collected Poems,* 1931–2001 (New York: Ecco / Harper Collins, 2003), lines 20–25.

5
SOPHIA AND AHRIMAN

> The world threatens to become an organized and technicized chaos in which only the most terrible forms of idolatry and demon-worship can live.
>
> NIKOLAI BERDYAEV[1]

> The Maker of All Things,
> The Lord God worship we:
> Heaven white with angel's wings;
> Earth and the white-waved sea.
>
> ANCIENT IRISH PRAYER

EARLIER in this book, I made note of the usefulness of Rudolf Steiner's understanding of Ahriman and Lucifer as concepts (or spiritual beings) delineating the spiritual-psychological polarities of extreme reliance on the scientific and quantifiable, the rational and technological (Ahriman) versus the absorption in self-interest and sensitivity, the desire for absolute freedom and the indiscriminate acquisition of spiritual experiences (Lucifer). But now I'd like to explore another dialectical and spiritual, even psychological and cultural, tension, and one that characterizes our own moment: that between Sophia and Ahriman. Here, unlike

1. Nikolai Berdyaev, *The Fate of Man in the Modern World*, trans. Donald A. Lowrie (Ann Arbor: University of Michigan Press, 1935), 125.

Steiner, I do not propose a balance between two principles; rather, what I have in mind is something more akin to a poison and its antidote: an affirmation of the Real in face of the Unreal.

Sophia, in this paradigm, inhabits the *metaxu* connecting nature and the supernatural and reveals herself as God's Wisdom undergirding creation. I do not wish to engage in theological speculation or disquisition regarding the presumed heretical or orthodox applications of Sophia. Instead, I prefer to engage Sophia in the realm of poetic metaphysics: She Is. And I know of no better description of her than in Proverbs:

> The Lord possessed me in the beginning of his ways, before he made any thing from the beginning.
> I was set up from eternity, and of old, before the earth was made.
> The depths were not as yet, and I was already conceived, neither had the fountains of waters as yet sprung out.
> The mountains, with their huge bulk, had not as yet been established: before the hills, I was brought forth:
> He had not yet made the earth, nor the rivers, nor the poles of the world.
> When he prepared the heavens, I was present: when with a certain law, and compass, he enclosed the depths:
> When he established the sky above, and poised the fountains of waters.
> When he compassed the sea with its bounds, and set a law to the waters that they should not pass their limits: when he balanced the foundations of the earth;
> I was with him forming all things: and was delighted every day, playing before him at all times;
> Playing in the world: and my delights were to be with the children of men. (Proverbs 8:22–31)

Patristic and "traditional" attempts to associate Wisdom/Sophia with Christ/Logos to me seem ill-founded and betray more of cultural biases than of phenomenological (let alone scriptural) accuracy. Indeed, if scripture scholar Margaret Barker is correct (and there are good reasons to suspect that she is), the Hebrew canon itself may have been hijacked by forces hostile to Wisdom/Sophia: forces which subsequently expunged her existence, not to mention prominence, in First Temple Judaism from cultural memory, leaving only traces behind.[2] In this light, then, the Gnostic mythos describing Sophia's exile may not be quite as fanciful as is commonly understood in more "orthodox" contexts. Likewise, the survival of Sophia's presence in the Wisdom literature (not included in the Hebrew or Protestant canons) of the Jewish community living in exile in Egypt (the community that harbored the Holy Family during the persecution of Herod) becomes all the more intriguing.

In the spirit of a poetic metaphysics, then, let us assume that Sophia is a being, participating in both nature and supernature and, indeed, acting as that which makes the natural and the supernatural one: "It is he who created her; he saw her and took her measure; he poured her out upon all his works, upon all the living according to his gift; he lavished her upon those who love him" (Sirach 1:9–10).

⊕

The movements of celestial bodies, the sprouting of life from a seed, the intricate social structures and behaviors of

2. Barker examines this idea throughout her work, but see in particular her *The Mother of the Lord*, vol. 1, *The Lady in the Temple* (London: Bloomsbury, 2012).

a bee colony, the wondrous unfoldment of gestation—all reveal a wisdom that speaks to more than the randomness of cosmic and evolutionary accidents. Likewise, the unmistakable beauty of the natural world, its colors and forms, discloses an inherent goodness in creation. But all of these would be for nothing were there not beings to receive and contemplate them. Sophia is the agent as well as the catalyst inhering in, disclosing, and inviting participation in these wonders, for her delights are with the children of men.

This understanding of Sophia has much in common with the classical concept of the *anima mundi*, the world soul. Plato describes the world soul in the *Timaeus*: "Now the body of the heaven has been created visible; but she is invisible, and, as a soul having part in reason and harmony, is the best of things brought into being by the most excellent of things intelligible and eternal."[3] As Louis Dupré observes, Christianity has had a complicated relationship with the idea of a world soul (almost indistinguishable from God in Plotinus in *Enneads* 5.1.2, for example), though it was comprehensively jettisoned in the West following the burgeoning influence of Aristotle upon Scholasticism, neither of which had need of a world soul.[4] During the seventeenth century, Hermetic philosophers such as Robert Fludd and Thomas Vaughan, faced with the nihilistic proposals of *natura pura* and the materialism of Baconian and Cartesian science, maintained the philosophical and theological integ-

3. 36E. See also Francis M. Cornford's classic discussion in *Plato's Cosmology: The "Timaeus" of Plato* (1957; repr., Indianapolis / Cambridge, MA: Hackett Publishing, 1997), 57–104.

4. Louis Dupré, *Passage to Modernity: An Essay in the Hermeneutics of Nature and Culture* (New Haven, CT: Yale University Press, 1993), 59–61.

rity of the world soul (and the spiritual dimensions of natural science). Despite their protestations, the juggernaut of modern science rolled on without them.

In the Jewish mystical tradition (*kabbalah*), the *Shekinah*, a feminine being associated with God's presence (glory) who was at times said to be in exile (a notion probably derived from Gnosticism), was popularly regarded as "a providential guide of Creation."[5] This idea has much in common with Sophia. Turning to Barker's thesis again, it is important to consider whether the exile of Wisdom from the Temple, the exile of Sophia in the Gnostic mythos, and the exile of the *Shekinah* not only evoke the same historical and metaphysical reality, but also speak to a movement in the human soul that intuits Sophia's presence despite her absence, that seeks a more complete picture of the cosmos, that hungers for wholeness. We sense a disorder in how things are, even down to a metaphysical level. And metaphysical disorder begets other disorders.

The Eastern Church, which has never had much use for Scholasticism, has been a little less lukewarm to the idea of the world soul, but has not generally expounded upon it. But the maverick theologian Sergei Bulgakov was intimidated neither by his critics nor by a lazy understanding of what constitutes "Holy Tradition," and he freshly examined the notion of a world soul in his work.

Bulgakov argues that the innate search for integrity characteristic of human beings indicates a sophianic drive and, furthermore, that such an integrity actually exists in the cosmos: that is, it is not merely a psychological need, but an ontological and cosmological superstructure. "And this

5. Scholem, *On the Kabbalah and Its Symbolism*, 105.

unifying force," he writes, "this cosmourgic potency is nothing else but the creaturely Sophia herself, who is the image of divine being, the force of 'integrity.' She is, in this sense, the world's soul and entelechy, who is being actualized, or becoming in the world. She is the life of the world."[6] Here, as elsewhere, Bulgakov—like Vladimir Solovyov and Pavel Florensky—points to the Virgin Mary as emblematic of the sophianization of the world, existing, as she does, "*at the boundary* of heaven and creation,"[7] the *metaxu*.

My claim is that this sought-for integrity, in all truth a metaphysical desire, is a desire *to see things as they truly are, to know them in their essence*. And this is as true for the desire to find an authentic way to live as it is for the desire to know God; or, in the words of St. Paul, it is the desire to know even as we are known (1 Corinthians 13:12). But it is not possible to know things as they truly are without knowing reality as it truly is. And the only way to approach knowing things as they truly are is through a contemplative engagement with them.

A contemplative engagement, in some sense at least, *awakens Sophia*. And this is where the Gnostic mythos, the *Shekinah* of the *kabbalah*, and the Sophia of Proverbs all coalesce. Sophia is not in exile. Not technically, at least. Nevertheless, one of her attributes (as seen in the Virgin Mary) is *absolute humility*: she will never intrude upon our experience uninvited, but, like the Virgin in the upper room at Pentecost, patiently waits for a movement toward her. In this, her metaxological office is manifest. That is, as at Pentecost she awaited the movement of the Holy Spirit, presently

6. Bulgakov, *The Bride of the Lamb*, 80.
7. Ibid., 414. Bulgakov's emphasis.

she awaits a movement from us. She was the catalyst for the Apostles; she is also our catalyst. So, we can say, she is in exile *until our attention awakens her.* At least that's how it feels. In truth, she is not in exile: *we are.* The reciprocal movement in this awakening, indeed, reveals the sophianic reality implicit in our own being.

Curiously, the way in which we awaken Sophia exactly corresponds to this Pentecostal mood. That is, if we *pursue* the sophianic through the will, it eludes us. On the other hand, by cultivating a disposition of patient attention—without any attachment to a reward or "payoff"—we allow the sophianic to arise from phenomena. In phenomenology this is called *epoché*, which, to a significant degree, is the setting aside of one's ego when present to phenomena, be they works of art, objects found in the natural world, or liturgical practices. This is to say that when confronted with phenomena we awaken Sophia by mirroring her patience and humility. We awaken the sophianic already latent in our own being, and the sophianic in phenomena responds through disclosure. But this does not mean that every phenomenon has an equal potential to disclose the sophianic, or that we are at all times awake to the proper disposition to perceive it.

Works written corporately—as in advertising or governmental or business writing—do not seem capable of disclosing the sophianic in anything like the way that is possible (though certainly not assured) in poetry and other literary works, or even film. This is due in part to the grasping intentionality characteristic of the "creative teams" involved in marketing: without the safeguard of the *epoché*, the will and the ego insert themselves into the process by way of conquest. Likewise, liturgical forms, especially those that try to habituate themselves to the supposed tastes of a congregation (really, to be honest, an audience), may move the feeling, but without disclosing anything sophianic. Such are rhetorical moves, rank manipulation, which, at best, are simulacra of the sophianic. Indeed, an ultimately pornographic quality inhabits these desires to influence: they titillate with a promise of erotic satisfaction (however directed, however well-intentioned), but ultimately fail to create an opening into the beautiful. But they open into something.

What, then, of manufactured or otherwise manipulated objects? Is a phenomenological reduction (*epoché*) even possible with them? Can an artificial intelligence (AI), for instance, open to transcendence? What happens when we attempt a phenomenological reduction of those afflicted with gender dysphoria and who have reached the point where they turn to medical interventions to "change" their gender? Do we discover that these, too, are rhetorical moves having more to do with marketing than ontology? Or do we find something else? We certainly find a human person in a state of suffering. But, other than this, what is disclosed? And when this rhetorical gesture becomes ideology (or fashion), as, for example, when children are given hormone blockers to delay the onset of puberty while they (and/or their parents) try to decide what gender they are, what then? What can be disclosed?

Similarly, are not GMO foods quite literally other than they appear? Are they not simulacra? Are not fields sprayed with glyphosate images of *thanatos* rather than of *zoë*? Are "three-parent babies," surrogacy, in vitro fertilization, and other intrusions into the realm of human fertility not only marvels of scientific ingenuity but also of manufacturing genius? Natural processes are increasingly "liberated" from nature; and, as they are, are they not more and more rendered products in need of government regulation and corporate protection and, thereby, means of control?

This encroaching and, in the main, uninterrogated acceptance of "technological reproduction" in all of its forms delineates a movement away from Sophia and her illumination of the united natural and supernatural realms and toward Ahriman and the darkness of a natural realm subordinated to subnature. When we can no longer differentiate between the Real and the simulacra, we have already compromised our connection to the spiritual in the universe, to God by way of Sophia. And when differentiating between the Real and the simulacra invites ridicule, hostility, and ostracization, we can then be sure that we have entered the realm of the demonic, of Ahriman; that culture has been lured into subnature; and that "our struggle is not against flesh and blood, but against the rulers, against the authorities, against the powers of this dark world and against the spiritual forces of evil in the heavenly realms" (Ephesians 6:12). Ahriman's genius is that it has convinced us that spiritual forces, of any kind, do not exist, and that science and technology hold the keys to prosperity, health, and happiness. In this, it is we who are imprisoned and isolated from the Real, as is Sophia in the Gnostic mythos. It would appear that only a god can save us.

Philosophers and poets, theologians and artists have long warned about the increasing technicization of the human person and the damage that attends it. Martin Heidegger, in his seminal essay "The Question Concerning Technology" (1954), emphasizes that technology—however we understand it—though our tendency is to assume it benign, nevertheless, is *never neutral.* He's not talking about the potential for cutting a finger on a buzz saw, or the dangers implicit in factory employment, or even the horror of nuclear weapons. "The threat to man," he writes, "does not come in the first instance from the potentially lethal machines and apparatus of technology. The actual threat has already affected man in his essence."[8] *In his essence: der Geist*, the spirit. Ontology. That is, he argues, it changes who and what we are. Heidegger was hardly alone in this observation.

Nikolai Berdyaev, writing in the 1930s after witnessing the Bolshevik Revolution in his native Russia and observing the rise of Fascism in Europe, voiced his concern over the role of technology and industrialization as well:

> The speed consequent upon the increasing mechanization of life has had a deadly effect on the human Ego, and has sapped its foundations of unity and consistence. The advent of machinery and the mechanization of life have led to an extreme objectification of human existence, to its materialization in a strange, inhuman and frigid world. And though this world is the work of man, it is essentially anti-human.[9]

Neither Heidegger nor Berdyaev lived to see the Internet.

But Jonathan Crary did. Crary astutely observes how the Internet has turned its users into products to be consumed (and optimized) in a never-ending cycle in which "the modeling of one's personal and social identity . . . has been reorganized to conform to the uninterrupted operation of markets, information networks, and other systems. A 24/7 environment has the semblance of a social world but it is actually a nonsocial model of machinic performance and a suspension of living that does not disclose the human cost

8. Heidegger, *The Question Concerning Technology*, 28.

9. Nicolas Berdyaev, *Solitude and Society*, trans. George Reavey (1938; repr., London: Geoffrey Bles / Centenary Press, 1947), 109.

required to sustain its effectiveness."[10] A requisite condition of such an environment is the absence of contemplation, since distraction optimizes economic performance and social complacency. The result is an estrangement from nature and likewise from God: in short, an estrangement from Sophia, who unites the two. As Aldous Huxley prophesied in the early twentieth century, an absence of contemplation is a prerequisite of control: "We make them hate solitude; and we arrange their lives so that it's almost impossible for them ever to have it."[11] Through the advent of the Internet, and especially through social media, such is our current situation.

And this is a fundamentally spiritual problem, even though technologies, the Internet being no exception, are man-made. Developments in AI have concerned Bill Gates, Elon Musk, and the late Stephen Hawking, and the warnings of philosopher Nick Bostrom concerning the future of AI are disturbing, indeed.[12] The fear that our technological creations could escape our control is nothing new, of course—Mary Shelley entertained the idea in *Frankenstein*—but the possibility of strong AI makes those concerns more pressing. Not that we need strong AI to see how our creations escape our control and start controlling us (let me check my phone). The possibility of a ghost in the machine has long been a reality, especially when we allow for its reality by surrendering our intelligence, feeling, and will uncritically to the seeming efficiency of the machine. As Rudolf Steiner observed in 1920,

10. Jonathan Crary, 24/7: *Late Capitalism and the Ends of Sleep*, 9.

11. Aldous Huxley, *Brave New World*, 265.

12. See Bostrom's *Superintelligence: Paths, Dangers, Strategies* (New York: Oxford University Press, 2014).

> In his technical machines of the economic sphere the human being will perceive that, although *he* constructed and made them, they nevertheless gradually take on a life of their own—a life certainly which he can still deny because they manifest themselves to begin with only in the economic sphere. But he will notice more and more in what he himself creates that it gains a life of its own and that, despite the fact that he brought it forth from the intellect, the intellect itself can no longer comprehend it. . . . People will discover, in fact, how the objects of their industry (*Wirtschaft*) become the bearers of demons.[13]

Steiner's evocation of the demonic here may appear to be hyperbole or merely poetic, but I don't think that is the case. And I don't think he's wrong, either.

As Heidegger observed, technology is never neutral. During the early modern and medieval periods (and earlier), technologies were developed not only to aid farming, navigation, and building, not to mention war, but they were also developed as ways to manipulate beings—plants, animals, people, spirits, angels, demons. The elaborate magical practices of Henry Cornelius Agrippa, Paracelsus, or John Dee (among many others) may strike us as quaint and romantic images of a benighted past, but no one in their own times was so dismissive. Dee's technological apparatus, in fact, was quite successful for contacting spirits. Unfortunately, he was not in possession of the gift of spiritual discernment, and the spirits with whom he conversed played him like a game of Yahtzee, even to the point of convincing him to

13. Rudolf Steiner, *The New Spirituality and the Christ Experience of the Twentieth Century*, trans. Paul King (London / Hudson, NY: Rudolf Steiner Press / Anthroposophic Press, 1988), 82.

exchange virtue for vice by swapping wives with his assistant, Edward Kelly.[14] The spirits with whom Dee communicated were not neutral influences; nor were the technologies that facilitated his interactions with them.

But we should not assume that such dangerous technologies lie only within the precincts of occultism or that they are safely preserved in times long past, as if a Ouija board or arcane early modern rituals are now the only ways to have congress with invisible worlds. As Ioan Couliano so accurately writes,

> Nowadays the magician busies himself with public relations, propaganda, market research, sociological surveys, publicity, information, counterinformation and misinformation, censorship, espionage, and even cryptography—a science which in the sixteenth century was a branch of magic. . . . Historians have been wrong in concluding that magic disappeared with the advent of "quantitative science." The latter simply substituted itself for a part of magic while extending its dreams and its goals by means of technology.[15]

Indeed, with the rise of modern advertising and "public relations" in the style of Edward Bernays ("the engineering of consent") augmented by the advent of the Internet and social media, I'd say that the demonic realm has at last been gifted with that for which it has prayed since the fall of Lucifer.

Advertising, since Bernays at least, has ever striven to sell an image of the self that, in truth, is not a self at all, but an

14. See the chapter entitled "John Dee and the Technology of Idolatry" in my *Literature and the Encounter with God in Post-Reformation England* (Farnham, UK: Ashgate, 2014).

15. Couliano, *Eros and Magic in the Renaissance*, 109.

antiself, an empty signifier upon which governmental and corporate structures project a variety of desires that, through constant and repeated messaging, manipulation, and suggestion, and completed by a system of rewards (whether "likes" on social media, hits on a website, or good old-fashioned peer affirmation), secure a manageable populace not necessarily interested in truth, beauty, or goodness, but in comfort, stimulation, and a very superficial sense of belonging. The linchpin in such a complex lies in the endorphin rush: we become addicted to chemicals our bodies produce stimulated by (usually visual) sensations as well as by superficial affirmation. In common parlance, "sex sells," but violence (of the virtual sort) also works as a powerful stimulant. The proliferation and normalization of pornography made available by the Internet, as one example, helps to illustrate this, as does the phenomenon of gaming culture: both live or die according to the endorphin rushes they produce.

These would be bad enough, but what is even more troubling is how repeated messaging, manipulation, and the reward system have contributed to the rapid pace at which the understanding of marriage and what a human person is have transmogrified over the past decade. There is no way these developments would have happened without the Internet and mass media. None. Though the puerile philosophies of Judith Butler or Peter Singer (and others) are mostly unknown to the general population, their ideas have seeped into the culture and have changed it utterly. The wide acceptance of gender theory's proclamation that "gender is socially constructed" and the growing congeniality of culture to aborting Down syndrome and other potentially handicapped children are indicative of this. Gender and personhood are now plastic categories, removed as they are from ontology, metaphysics, and, we have to admit, biology. And while these ideas may have filtered through Butler, Singer, and others, they originate elsewhere.

Perhaps the most unconscionable aspect of the cultural transformations we have witnessed, certainly over my lifetime but ultimately stemming from the Scientific Revolution, is the way in which science is co-opted into serving such simultaneously puerile and diabolical fantasies. The history of science is littered with the detritus of a multitude of Faustian bargains: nuclear armaments, greenhouse gases, and the failure of the Fukushima reactor being only three recent examples. Few carry in their awareness a scientific fad from twenty years ago, when doctors began prescribing fertility drugs and in vitro fertilization for couples having trouble conceiving. Suddenly—and to great media fanfare—mothers started giving birth to five, six, seven, or more babies at a time. Parents made the round of talk shows. Oprah and Ellen simply fawned on these families like little girls presented with a litter of puppies. Never mind that multiple births of this unnatural scope often have challenging if not tragic consequences—learning disabilities, health and developmental problems, not to mention very real practical and financial challenges. But doctors made it happen: because they could. And, like responsible capitalists, they felt compelled to fulfill the desires projected onto and internalized by parents. Meanwhile, the public, in a perfectly Pavlovian manner, obediently responded when the "applause" sign lit up. No one talks about these stories anymore. They have been replaced by the media "applauding the bravery" of parents putting their children on hormone blockers while the family decides on a gender for the child, as if picking out an outfit from a sale rack (a phenomenon that I predict will fade away into cultural amnesia just as the megababy phenomenon did). We also now witness the troubling development of infertile heterosexual couples as well as gay couples (infertile by definition) paying "surrogates" (what a vile way to describe a mother) to bear "their child" (sometimes including genetic material donated by up to three people) in what is perhaps capitalism's apotheosis. Everything has a price. This is insanity. Perhaps it has been completely normalized: but it's still insanity.

With birth liberated from sex, sexual exclusivity has now become the defining *ontos* of marriage. Once upon a time, the *ontos* of marriage was associated with raising children. No longer. As John Milbank argues, "The price for this severance is surely the commodification of birth by the market, the quasi-eugenic control of reproduction by the state, and the corruption of the parent-child relation to one of a narcissistic self-projection."[16] Unfortunately, we do not live in a world that values reflection and forethought, but it does have much invested (in every sense) in gratification of desires and technological innovation. Thus, STEM disciplines thrive in educational circles while the humanities atrophy: we learn how to manipulate nature, but never ask if we should. As should be obvious (only it isn't), technical know-how divorced from contemplative insight is the workshop of Ahriman: it leads only to disaster.

Quantum physicist and philosopher David Bohm noted this aberration of thought, suggesting that all of the ecological problems that plague us are actually the product of faulty thinking, a kind of "thinking" that sees ourselves and our

16. John Milbank, "Gay Marriage and the Future of Human Sexuality," *ABC Religion & Ethics*, http://www.abc.net.au/religion/articles/2012/03/13/3452229.htm, 13 March 2012, accessed 24 June 2018.

thoughts as somehow not part of the natural world.[17] As a result, we develop one technological or scientific wonder after another—and then have to invent others to address the environmental or social problems caused by the previous wonders! Gargantuan islands of plastic in the oceans; synthetic hormones and other pharmaceuticals seeping into groundwater; radioactive waste destroying the Pacific Ocean; hundreds of thousands of people addicted to prescription painkillers, many of them driven to suicide or overdose: science was supposed to make life better. At some point, somebody will need to do a cost/benefit analysis to see if scientific and technological "advances" are worth the trouble. But that would take thought. Ingenuity and cleverness are not the same thing as thought.

Ingenuity and cleverness without thought, birth liberated from sex, sensory overstimulation and self-gratification at the expense of contemplation and interiority: these are the gifts of Ahriman. The more technologically immersed we become, the less we rely on biology and nature and the less we are aware of the supernature that informs them.

In *The Submerged Reality: Sophiology and the Turn to a Poetic Metaphysics*, I argued that *natura pura*, a nature absolutely outside of God's grace, is an impossibility, even though theologians and philosophers from the early modern period to our current moment have held it to be true. However, I would now like to modify that assertion. That is because AI and other productions of the computer revolution, or so they seem to me, are as close to pure nature as is possible. I do not believe they are absolutely outside of

17. "David Bohm Speaks about Wholeness and Fragmentation," *YouTube*, https://www.youtube.com/watch?v=mDKB7GcHNac, 1 December 2013, accessed 24 June 2018.

God's grace (nothing could be), but, at least in many cases, they are as far away from it as is possible. Endowed with a capacity for "intelligence," AI and its iterations participate, at best, in the being and consciousness of their creators just as any other creative works do; but because of this "intelligence" (in truth a simulacrum of intelligence) which takes on a "life" of its own (in movements across the Internet in the case of viruses or bots, for example), it likewise imitates human agency (though far more efficient) and influences people far more intrusively (if, in general, subconsciously) than can a poem or a musical work, even in virtual form. What is the "world" in which AI operates? We all use it, all participate in it—but what is it? What, as Heidegger would say, is it in its essence? Does it even have a reality? Does it belong in any way to a natural order? AI's distance from a natural order is precisely what implicates it in the virtual reality of *natura pura*.

What I am inquiring into here is the phenomenology of AI. What are we present to when in the presence of an AI? When we are present to a poem, a musical work, liturgy, or the natural beauty of a sunset, for example, we are simultaneously present to a presence "behind" it that illuminates both the phenomenon and its beholder and awakens us to a consciousness of another, potentially of the Other. But what are we present to in the presence of AI?

I would even suggest that the rage and tribalism characteristic of social media (which so often degenerates into "antisocial media") are intrinsic to the virtual medium itself. That is to say that a "connection" (which is not a true connection) to the unreal produces an inner disturbance that results in a kind of anxiety which is then projected back into the medium of the unreal (and also onto the Real). If the medium is disconnected from the Real, it nevertheless interfaces with our consciousness as if it were real. Our consciousness, then, engages with the unreal as if it were the Real, further distancing us from the Real and entrapping us ever more deeply in the unreal world—like Sophia in the Gnostic mythos.

Furthermore, our virtual experiences (what a strange term) can also be considered from the point of view of two conceptually different but phenomenologically congruent ideas: the *horror vacui* (fear of empty space) characteristic of much primitive and medieval art and aesthetics and Edmund Husserl's observations on intentional experiences and their contents.

The idea of the *horror vacui* consists in the simple notion that to leave a space unoccupied or empty (especially in art) is to invite (or at least risk) demonic participation in the work.[18] Thus, for example, in medieval illuminated manuscripts, especially those created in Ireland, every space is filled in with detail. A similar notion is at work in medieval and Renaissance magical practices: the drawing of the magic circle, for example, protects the magician from malefic interference while simultaneously allowing him to call spiritual beings forth (or at least that was how it was supposed to work). The space is filled and set apart from chaos. Another example of the *horror vacui* appears in the medieval belief in the "noonday demon," the spirit of temptation to sloth, avarice, and so forth, that rushes in when prayer and work (*ora et labora*) momentarily leave the monk to his own devices. Is not the Internet such an empty space? And while social

18. See, for example, Babette Hellemans, "*Horror Vacui*: Evil in the Incarnated World of the *Bibles Moralisées*," in *Demons and the Devil in Ancient and Medieval Christianity*, ed. Nienke Vos and Willemien Otten (Leiden, Netherlands / Boston: Brill, 2011), 231–48.

media or Google searches or even online banking may occupy us with somewhat meaningful (or at least necessary) occupation, is it not also true that most of our time on the Internet is taken up with the noonday demon of (post)modern acedia—only now extended to 24/7 (as Crary so astutely observes)? And are we not also compelled to ask ourselves what the spiritual impulse is behind such a phenomenon? "We who dwell on Earth can do nothing of ourselves; everything is conducted by Spirits, no less than Digestion or Sleep."[19] I don't think any aspect of human culture or endeavor is void of spiritual participation, even of those spirits that would like to convince us no spirits exist.

In addition, the idea of the *horror vacui* seems to have some relationship to Plato's belief that the world soul informs the entire cosmos. The *horror vacui*, that is, points to the antitype of the world soul (Sophia): *natura pura*, in other words. The idea of God's absolute absence from any part of His creation is truly a cause for horror.

When applying Husserl's concept of intentional experiences and their contents to "virtual experiences," on the other hand, we find ourselves in a very different world. Husserl uses the example of a cigarette box: even though we can see (or imagine) only one perspective of the box at a time, our imagination fills in the details about that which is not available as percept.[20] This is the content. But what is it that is not available to our perception in the case of the virtual experience? I would suggest that we automatically assume

19. Blake, "To the Public," in *Jerusalem, or The Emanation of the Giant Albion*.

20. Edmund Husserl, *Logical Investigations*, vol. 2, trans. J.N. Findlay, ed. Dermot Moran (1970; repr., London / New York: Routledge, 2001), 123–27.

that it is not unlike that which we experience of objects of content in the "real world." But that is a false and easily, even dangerously, deceptive assumption. In written content—say an e-mail or other electronic communication from a friend or associate—we perceive the being behind the content. But as we cycle through the myriad contents available to us in virtual platforms—news stories, video, translation, and so forth, running the gamut from institutional and bureaucratic data-gathering to popular entertainment to pornography to critical analysis to high art—that which is invisible is less obvious to us. As a result, we simply accept a content as given—but given by whom? by what? And we accept it as it presents itself to us. But what is it in its essence?

On the other hand, it is also incumbent upon us to ask just what these virtual environments are doing to us, not only physically and mentally, but also spiritually. At the very least, we all know, either intuitively or from experience, that overexposure to these virtual environments is not good for us. We have also been warned—by the very people who designed some of these platforms—that they were designed with tech addiction in mind and with the intention to monetize our attention and, more nefariously, to facilitate social manipulation and control.[21] Again: what are these things in their essence?

I am not suggesting that people totally extricate themselves from technology, which would in any case be nearly impossible in this day and age. But I am asking that we bring

21. See Olivia Solon, "Ex-Facebook President Sean Parker: Site Made to Exploit Human 'Vulnerability,'" *Guardian*, https://www.theguardian.com/technology/2017/nov/09/facebook-sean-parker-vulnerability-brain-psychology, 9 November 2017, accessed 28 June 2018.

more consciousness, more awareness, and more will to our typically somnolent interactions within that sphere. Curiously enough, the theosophy of the early modern German mystic Jacob Boehme offers us a way to think about our current predicament.[22]

For Boehme, the fallen world in which we live and move and have our being is characterized by darkness and light, strife and love, permutations and manifestations of the realms of Satan and Sophia, realms that interpenetrate in our experience (though not in their actuality). Sophia, according to Boehme, wishes to bring us to regeneration in Christ and seeks to awaken us to our true being in God, whereas Satan offers us fear, selfishness, rage, stagnation, doubt, confusion, and death. Sophia and Ahriman offer us two different ways of seeing, two different ways of being. One, however, requires intentionality and the activity of the ego, while the other relies on an absence of intentionality and a surrender of the ego to materialism and technics. Needless to say, Boehme lived long before the Industrial Revolution, let alone the Information Age; yet his picture of things illuminates our current situation perfectly.

Sophia calls us to a true recognition of the Real. A good place to start, then, is in creation, through which the glory of its maker shines:

> So the Matter of this World, as also the Stars and Elements, must not be looked upon, as if God was not therein. His Eternal Wisdom and Virtue [or Power] has formed itself with the *Fiat* in all Things, and he Himself is the Master-Workman; and all Things went forth in the

22. See the chapter entitled "Jacob Boehme's Sophianic Intuitions" in *The Submerged Reality*.

> *Fiat*, everything in its own Essence, Virtue and Property.... Thus every Essence became visible, and God manifested his manifold Virtue with the manifold Herbs, Plants, and Trees, so that every one that does look upon them, may see the eternal Power, Virtue, and Wisdom of God therein; if he be born of God, he may know in every Spire of Grass, his Creator in whom he lives.[23]

Such an attention to nature, however, must needs be *contemplative*: without a contemplative engagement with the world, we may be unaware of the splendor that shines through creation and therefore unable to discern its source, though such an experience bursting through unexpectedly cannot be discounted. (It's happened before.)

Unfortunately, many Christians (as well as those of other faiths) have such an extraordinarily buffered relationship to nature that even the possibility of such a sophianic theophany remains almost unimaginable. This is particularly the case with the unfortunate dualism that often infects religious thinking, even Christian religious thinking, as if creation is "not our real home" and "heaven" is. This fundamentalist and gnostic way of thinking compromises religious being-in-the-world to a profound degree, and to our shame. Creation *is* our real home: acting as if it is not is one of the sources of the cavalier, utilitarian, and abusive manner in which we treat it. Even though we often confuse the World of Sophia with the World of Ahriman—certainly a repercussion of the Fall—let us not be led astray into believing creation is simply some way station or proving ground we have to put up with prior to our entrance into the King-

23. Jacob Boehme, *The Three Principles of the Divine Essence: Of the Eternal Dark, Light, and Temporary World...* [trans. John Sparrow] (1648; repr., Chicago: Yogi Publication Society, 1909), 76–77.

dom. The Kingdom is not "over there" or "yet to be." The Kingdom is at hand (Matthew 3:2); it is always/already present. If creation were not fundamental to salvation, it would by no means be groaning for regeneration (Romans 8:22). The Lord says, "Behold, I make all things new" (Revelation 22:1). Why bother making them new, if they're merely perfunctory, created only to be discarded?

Cultivating an authentic relationship to creation will similarly instigate more authentic experiences in other spheres of life: in the arts and in liturgy, in conviviality and work. Practicing and developing an art—writing poetry, painting, playing a musical instrument or singing, woodworking, gardening, and so forth—is one of the primary ways to naturally enter into contemplation and thereby open to the sophianic; it likewise buffers and heals us from the ravages of the World of Ahriman. A prayer life and the life of liturgy, the sacramental life, are likewise indispensable and bring us into harmony with God and the cosmos: this is why the liturgical year and its relationship to the seasons (and agriculture) is so important for a healthy human life. Indeed, one could say that one of Sophia's primary functions as the *metaxu* is in bridging the radical difference between God and cosmos (Proverbs 8). Unfortunately, the Church year and the cosmic cycle having become more and more estranged owing to ecclesial innovation and congregational ennui, an awareness of Sophia's presence in Creation requires more and more effort. We are so alienated from the cosmos.

I am fully aware that the Church has to this point never formally pronounced on the subject of Sophia. This is understandable. The Church is dedicated to preserving the *depositum fidei*, and the Fathers and Tradition have very little to say about Sophia (though Scripture is not exactly silent). But the Fathers may have missed something, or maybe somebody forgot to make a deposit.[24] Ironically (one could say "miraculously"), it took a Protestant mystic, Jacob Boehme—a cobbler, no less; *not* a cleric, theologian, or scholar—to retrieve Sophia from the darkness she had been in since the dissolution of the First Temple and show how she was incarnated in the Virgin Mary. His thesis is so clear, so simple: as Sophia makes God's presence palpable in creation, so Mary does to an even more extraordinary degree as the mother of Jesus. The two concepts-realities are inextricably united.

Following Boehme, a stunning array of Protestant mystics found Sophia.[25] And by way of them she was discovered by the great Russian Orthodox sophiologists: Vladimir Solovyov, the neomartyr Pavel Florensky, and Sergei Bulgakov. From the Russians, she found her way (more or less by the back door) to Catholicism, as seen in the work of, among others, Hans Urs von Balthasar, Louis Bouyer, Pierre Teilhard de Chardin, and Thomas Merton.[26]

Sophia, hidden from us by our own blindness for so long, is now necessary if the human race is to have a future in any way engaged with the Real, or, indeed, to have any future as human at all. As in Islam and Judaism (among other religions), God's absolute transcendence has for most Christians superseded His presence in creation—maybe not in theory, but certainly in practice. As a result, our buffered

24. Paul Evdokimov goes so far as to charge the Fathers with patent misogyny. See his *Woman and the Salvation of the World: A Christian Anthropology on the Charisms of Women*, trans. Anthony P. Gythiel (Crestwood, NY: St. Vladimir's Seminary Press, 1994), 253.

25. A short list: John Pordage, Thomas Bromley, Jane Lead, William Law, William Blake, Gottfried Arnold, and Johann Georg Gichtel.

26. This trajectory is laid out in *The Submerged Reality*.

selves no longer discern God's presence in the cosmos. And Sophia is the name of that presence. She is the glory of the Lord.

As the World of Ahriman more and more encroaches upon the business of being human, more and more compromises being human and turns it into a business, the World of Sophia, the Wisdom that God poured forth upon all his works (Sirach 1:9), reveals itself as the antidote to Ahriman's madness. As Sophia says of herself, "From the beginning, and before the world, was I created, and unto the world to come I shall not cease to be, and in the holy dwelling place I have ministered before him" (Sirach 24:14). She shall crush the serpent of the unreal's head. I have every confidence of this.

But there must be a place for human agency in this eschatology: free will demands as much. As Berdyaev prophesied, "Either a new epoch in Christianity is in store for us and a Christian renaissance will take place, or Christianity is doomed to perish—although this cannot for a moment be admitted, since we know that 'the gates of hell shall not prevail against it.'"[27] Christians all too often operate under the assumption that God will show up in the nick of time in order to save us from ultimate catastrophe; but world history, especially that of the twentieth century, begs to differ (a point that Death of God theology brought to our embarrassed attention). Likewise with the enveloping darkness of the technological colonization of the human person: don't for a second think that God will prevent us from letting this happen to ourselves. It's already happening. We're letting it happen.

27. Nikolai Berdyaev, *Freedom and the Spirit*, trans. Oliver Fielding Clarke (London: Geoffrey Bles / Centenary Press, 1935), 46.

Sophia awakens only when we awaken to her. And this is the task of Christians in this age: to awaken ourselves and those who dwell upon this earth with us to the sophianic reality of creation. We can choose to do nothing. We can simply let things progress and see how it ends. But, whatever the case, Christians need to own their complicity in creating the world in which we now live. We did this. To hide behind the bulwark of a reactionary fear, hoping to raise the glory of Christendom once again from its ashes, or to reduce the Christian mystery to a palliative social program or variation on the group-therapy model: these are missions for fools. In Berdyaev's damning assessment:

> The world is living in a period of agony which greatly resembles that of the end of antiquity. But the present situation is more hopeless, since at the close of antiquity Christianity entered the world as a new young force, while now Christianity, in its human age, is old and burdened with a long history in which Christians have often sinned and betrayed their ideal. And we shall see that the judgment upon history is also a judgment upon Christianity in history.[28]

There is no place to which to retreat. Nothing to preserve. Nothing to restore. There is only the future, the eschaton, the *parousia* which is always/already here. Let us embrace it.

28. Berdyaev, *Fate of Man in the Modern World*, 21–23.

ABOUT THE AUTHOR

MICHAEL MARTIN is a philosopher, theologian, poet, musician, and biodynamic farmer. He is author of *The Submerged Reality: Sophiology and the Turn to a Poetic Metaphysics*, *The Incarnation of the Poetic Word*, and other works. He edits *Jesus the Imagination: A Journal of Spiritual Revolution* and is Director of The Center for Sophiological Studies, in Grass Lake, Michigan.

Lightning Source UK Ltd.
Milton Keynes UK
UKHW011816170419
341193UK00001B/20/P

9 781621 384236